Life Without a Designer

Evolution, DNA, and the End of Divine Design

(Book 4)

Dr Prabhat Das

MD, PhD, FACP

No God Required

How Science Explains Reality Without the Divine

Hardcover ISBN: 979-8-951067-09-8

Paperback ISBN: 979-8-951067-08-1

eBook ISBN: 979-8-951067-07-4

Printed in the United States of America

TABLE OF CONTENTS:

Author's Note

Why I Wrote This Series

Some books begin in libraries. Some begin in classrooms. This series began much earlier, in the mind of a curious boy growing up in India.

I was born into a Hindu religious family. Religion was not a distant theory in my childhood. It was part of daily life. It was in prayers, rituals, festivals, stories, customs, family conversations, and the quiet assumptions that surrounded me. God was not discussed as a possibility. God was treated as reality.

My father was a firm believer in an omnipotent God. To him, God was real, powerful, protective, and deeply involved in human life. His belief was sincere. It was part of his strength, his discipline, and his view of the world. I grew up respecting that seriousness.

But from my teenage years, another force began to grow inside me. Curiosity.

I wanted to know why things happened. I wanted to understand nature, disease, death, the stars, the human body, and the mind. I was not satisfied with an answer simply because it was old, sacred, or repeated by many people. I wanted to know whether it was true.

In high school, I became deeply impressed by the lives and writings of great scientists. I still remember going to the local library and borrowing books written by, or about, Newton, Darwin, Einstein, and others. Those books were not popular among students at that time. In fact, I may have been the first student to borrow some of them. They were covered with dust. Some had even been damaged by insects.

But to me, they were treasures. Those dusty books opened a new world.

Newton showed me that nature follows laws.

Darwin showed me that life has a history far deeper than ancient stories.

Einstein showed me that reality is stranger, deeper, and more beautiful than ordinary imagination.

These thinkers did not ask me to believe blindly. They invited me to understand. They did not frighten me with punishment. They opened the door to wonder. They showed me that truth does not become weaker when questioned. It becomes stronger.

That invitation changed my life.

My curiosity did not end with school. It continued through medical college, scientific training, clinical practice, and decades of work as a physician. In medicine, I learned the value of evidence. A diagnosis cannot be accepted because it feels comforting. A treatment cannot be trusted because it is traditional. A belief about the body must face reality. It must face observation, testing, correction, and outcome.

That same habit of mind gradually shaped the way I looked at religion & God.

Even now, at retirement age, I continue to read almost every day. Science, philosophy, history, religion, psychology, evolution, cosmology, neuroscience, and the human mind still fascinate me. The more I read, the more I see how far human knowledge has come. Many things once explained by God are now understood through natural causes. Thunder, disease, eclipse, famine, fertility, mental illness, death, life, and the vast universe itself no longer need supernatural explanation.

This book series, "**No God Required: How Science Explains Reality Without the Divine**", **is a 7 book series**, and is my attempt to share that long journey in clear and simple language.

I do not write these books to insult believers. Many believers are kind, sincere, moral, generous, and deeply thoughtful people. I know this from my own family, my own culture, my patients, my friends, and my life experience. Faith often gives people comfort, hope, discipline, community, and emotional strength.

But faith does not decide truth.

A belief may comfort us and still need examination. A tradition may be ancient and still be mistaken. A prayer may bring peace and still not prove that anyone is listening. A sacred story may be beautiful and still be human-made.

My goal is to understand how belief began, why it became powerful, and why it still feels so real to so many people.

Why did human beings create gods?

Why did religion become one of the strongest forces in history?

Why do beliefs learned in childhood feel like truth?

Why do different cultures create different gods?

Why did science gradually make divine explanations unnecessary?

Can human beings live meaningful, moral, compassionate, and courageous lives without God?

These are serious questions. They deserve serious answers.

This series is based on knowledge gathered from many great books, scholars, scientists, historians, philosophers, psychologists, and thinkers. I have tried to bring their insights together in language that an ordinary educated reader can understand. My hope is that these books will encourage readers to think more rationally, question inherited beliefs, respect evidence, and study these subjects more deeply.

I also hope they will show that life without God does not have to be empty, cold, selfish, or immoral.

A life guided by reason, compassion, truth, science, justice, and human responsibility can be deeply meaningful. We do not need divine command to care for one another. We do not need fear of hell to be moral. We do not need supernatural stories to feel wonder before the universe.

Human beings are already capable of love, courage, kindness, beauty, and truth.

The universe is already astonishing.

Life is already precious.

Our responsibility to one another is already urgent.

No God is required for that.

Introduction: Life Is Beautiful, But It Is Not Perfect

The First Breath

There are moments when life feels almost sacred.

A newborn takes the first breath. For a few seconds, the room becomes still. The mother waits. The father waits. The nurse watches. The physician listens. Then the tiny chest rises, the lungs open, and a cry enters the world.

Even a skeptical mind can feel wonder in that moment.

A small heart is beating. Blood is moving. Cells are using oxygen. The brain is waking into sensation. The body, so fragile and small, is already doing thousands of things no machine can fully imitate.

It is not difficult to understand why human beings have looked at life and seen the hand of God.

A bird feeding its young, a flower opening toward light, a spider weaving its web, a child learning to speak, a wound slowly healing, an eye receiving color and shape from the world. These things can move us deeply. They can make life feel planned, guided, and lovingly arranged.

But awe is not evidence.

Wonder is a human response. It is not a scientific explanation.

This book begins with a simple question:

Does life truly show scientific evidence of divine design?

Many people assume that it does. They look at the beauty and complexity of life and feel that God must be behind it. That feeling is understandable. But science cannot stop at feeling. Science must ask what the evidence actually shows.

When we look at life from a distance, it may appear designed. When we look closely, with the tools of biology, genetics, medicine, fossils, and evolution, a different picture appears.

Life is beautiful, but it is not perfect.

Life is complex, but it is not planned.

Life is astonishing, but it is not scientific evidence for God.

Wonder Should Lead to Questions

Wonder is not the enemy of science.

In fact, science often begins with wonder. A child looks at the stars and asks why they shine. A patient asks why the body becomes ill. A naturalist watches birds and asks why their beaks differ. A physician sees the body heal and asks how such repair is possible.

The problem begins when wonder is stopped too early.

Religion often turns wonder into a final answer.

The flower is beautiful, therefore God.

The eye is complex, therefore God.

The child is precious, therefore God.

The body is remarkable, therefore God.

Science takes a different path.

It does not say, "This is beautiful, so stop asking."

It says, "This is beautiful, so let us understand it."

That difference matters.

A rainbow is not less beautiful because we understand light. A flower is not less beautiful because we understand pollination. A newborn is not less precious because we understand oxygen, blood flow, genetics, and development. The eye is not less impressive because we understand evolution.

Understanding does not destroy wonder.

It makes wonder more honest.

The Illusion of Design

Human beings naturally see purpose.

We see faces in clouds. We find meaning in coincidence. We ask why something happened, even when there may be no personal reason at all. This habit may have helped our ancestors survive. If something moved in the grass, it was safer to assume there might be an animal there than to ignore it.

The mind that quickly detected agency had an advantage.

But the same habit can mislead us.

When we see a wing, we imagine it was made for flying. When we see an eye, we imagine it was made for seeing. When we see flowers attracting insects, we imagine someone arranged the relationship. When we see the human body breathe, heal, think, and reproduce, we may feel that such a system must have been designed by a higher intelligence.

That feeling is powerful.

But it is not enough.

The central claim of this book is clear:

Living beings look designed because evolution can produce the appearance of design without a designer.

Natural selection can preserve useful traits. Inheritance can carry them forward. Variation can supply new possibilities. Deep time can allow small changes to accumulate. DNA can preserve the record of ancestry. Fossils can show life changing across vast ages.

This is why the old argument from design becomes weaker as biology becomes clearer.

Religion says, “Look how wonderful life is. It must have been created by God.”

Science replies, “Look more closely.”

Life Is Not a Perfect Machine

If life had been created from nothing by a perfect, all-knowing, all-powerful, loving designer, we might reasonably expect a different living world.

We might expect bodies that are safer and more efficient.

We might expect no harmful genetic disorders.

We might expect no childhood cancers.

We might expect no parasites that survive by invading other bodies.

We might expect no dangerous childbirth.

We might expect no blind spot in the eye.

We might expect no appendix that can become inflamed and rupture.

We might expect no immune system that sometimes attacks the very body it is supposed to defend.

We might expect no world in which one living creature must tear apart another living creature in order to survive.

We might expect no mass extinction of most species that ever existed.

But that is not the world we see.

We see life that works, but not perfectly.

We see organs that are useful, but also vulnerable.

We see genes that carry information, but also errors.

We see immune systems that protect, but sometimes harm.

We see cells that divide to repair the body, but sometimes become cancer.

We see species that adapt, flourish, struggle, and disappear.

We see beauty and suffering woven together.

This does not look like perfect divine engineering.

It looks like evolution.

The Body Through a Physician's Eyes

The human body is remarkable. No honest physician can deny that.

The heart beats with quiet persistence. The lungs exchange gases every minute. The kidneys regulate fluid, salt, and acid. The liver performs complex chemical work. The skin protects. Blood clots when a vessel is injured. Bones heal. The immune system recognizes many dangers. The brain creates memory, thought, language, fear, hope, love, and imagination.

But a physician also sees the other side.

The heart can fail.

The lungs can fill with fluid.

The kidneys can shut down.

The liver can scar.

The immune system can attack the body.

Cells can become malignant.

Blood can clot in the wrong place.

A child can be born with a malformed heart.

A young person can develop leukemia.

An elderly person can lose memory, balance, dignity, and independence.

The body is not a flawless divine machine.

It is a fragile biological system.

This does not make the body less worthy of respect. It makes it more precious. But it also makes the design argument less convincing.

A perfect designer could have done better.

Evolution explains why the body is powerful but imperfect. It was not built from a clean blueprint. It was shaped by inheritance, mutation, selection, trade-offs, and survival.

Evolution Does Not Destroy Meaning

Some people fear that evolution makes life meaningless.

They worry that if human beings were not specially created by God, then we are nothing more than animals, accidents, or machines. They fear that without divine design, dignity disappears.

That fear is understandable.

But it is mistaken.

Evolution does not make life meaningless. It makes life connected.

It tells us that every living being belongs to a vast biological history. The tree outside the window, the bird in the sky, the fish in the sea, the dog sleeping beside a family, the bacteria in the soil, and the child learning to walk are all part of one living story.

We are not strangers dropped into nature from outside.

We are nature.

Our bodies carry the memory of earlier life. Our bones, lungs, hands, nerves, blood, eyes, and brains are not separate divine inventions. They are modified inheritances from a long natural past.

This does not reduce human dignity.

It gives dignity a natural foundation.

We matter because we can suffer, love, think, remember, cooperate, create, and care. We matter because each conscious life is brief and irreplaceable. We matter because our choices affect other living beings.

We do not need special creation to have value.

We need compassion, reason, and responsibility.

Evolution and the Origin of Life Are Different Questions

One common confusion must be removed at the beginning.

Evolution and the origin of life are related, but they are not the same question.

Evolution explains how living populations change after life already exists. It explains adaptation, diversity, common ancestry, speciation, extinction, and the branching history of life.

The origin of life asks a different question:

How did nonliving chemistry become the first life-like systems?

That question is still being studied. Scientists investigate prebiotic chemistry, RNA, simple membranes, mineral surfaces, hydrothermal vents, hot springs, energy gradients, wet-dry cycles, and organic molecules from space. Science does not yet have every detail of the first living system.

But not knowing every detail is not evidence for God.

A gap in knowledge is not a miracle.

History teaches this lesson again and again. Thunder was once explained by gods. Disease was once explained by spirits. Mental illness was once explained by demons. The movement of planets was once explained by divine intention. One by one, science replaced supernatural guesses with natural explanations.

The honest answer to an unanswered question is not "therefore God."

The honest answer is, "We continue to investigate."

That is the difference between science and religion. Religion often turns mystery into certainty. Science turns mystery into inquiry.

What This Book Will Show

This book will examine the major biological reasons why life does not require a divine designer.

First, we will ask what real evidence of a designer would look like. This is necessary because many people confuse appearance with proof. Life may look designed, but science must ask whether the evidence truly supports that claim.

Then we will examine why life looks designed. We will see how natural selection can create the appearance of purpose without any mind behind it.

We will look at the world before Darwin, when creation stories felt reasonable because people lacked better explanations. Then we will see how Darwin changed biology by showing that species could arise through natural processes.

We will examine natural selection, not as blind random accident, but as a powerful process in which inherited variation is filtered by survival and reproduction.

We will see evolution happening in real time, in bacteria, viruses, insects, plants, and other living populations.

We will enter deep time and the fossil record, where the history of life appears not as sudden divine creation, but as change, branching, extinction, and transformation.

We will consider the origin of life honestly, without pretending that science has every answer and without pretending that every unknown is evidence for God.

We will examine DNA, not as a magical code from heaven, but as a biological archive of common ancestry.

We will look at the tree of life, which shows that living beings are not separate divine products, but relatives.

Then we will look at the body itself.

The human body is impressive, but it is not perfectly designed. It carries traces of evolution: awkward compromises, vulnerable organs, painful weaknesses, and structures shaped by history rather than ideal engineering.

Disease will be especially important. Cancer, genetic disorders, autoimmune disease, infections, birth defects, aging, pain, and mental illness are not small problems for the claim of perfect divine design. They are central problems.

We will also look beyond human disease into the wider living world: parasites, predators, starvation, reproductive waste, animal suffering, and extinction.

Finally, we will examine human evolution. We will see that human beings were not specially created outside nature. We are evolved primates, part of the same living world we study.

This truth does not make us worthless.

It makes us honest.

Why the Design Argument Fails

The design argument is emotionally powerful because it begins with something real.

Life is amazing.

But the conclusion does not follow.

A snowflake is beautiful, but it is not carved by angels. A rainbow is beautiful, but it is not painted by a god. A crystal has order, but it does not require a mind. A river can carve a canyon without knowing what it is doing. Gravity can gather matter into stars without intention.

Natural processes can create structure.

Biology adds another powerful process: natural selection.

Natural selection can shape useful complexity over time. It has no mind, but it has consequences. It has no plan, but it filters. It has no mercy, but it preserves what survives and reproduces.

The mistake of the design argument is to assume that complexity always requires conscious design.

It does not.

A supernatural designer explains less than people think. Saying "God designed life" does not explain why bodies are imperfect. It does not explain why genes fail. It does not explain why children suffer. It does not explain why parasites exist. It does not explain why most species that ever lived are extinct. It does not explain why humans share anatomy and DNA with other animals.

The word "God" often stops the question just when the question is becoming important.

Science does not stop there.

Science asks how.

The Purpose of This Book

This book is not written in anger.

It is written with conviction.

It is not meant to insult religious people. Many believers are sincere, kind, intelligent, and morally serious. The problem is not the sincerity of the believer. The problem is the weakness of the explanation.

When religion claims that life proves God, that claim must be examined.

When religion claims that the human body proves divine design, that claim must be compared with disease, pain, birth defects, cancer, aging, and death.

When religion claims that humans were specially created, that claim must be compared with fossils, DNA, anatomy, and human evolution.

When religion claims that God is needed to explain life, that claim must be compared with what modern science actually shows.

The goal of this book is calm clarity.

Not ridicule.

Not hatred.

Not blind disbelief.

Only evidence.

And the evidence points in one direction:

Life does not need a designer.

Chapter Summary

Life inspires wonder, but wonder is not scientific evidence.

The appearance of design is powerful, but natural selection can explain that appearance.

Evolution explains the diversity and adaptation of life after life exists.

Origin-of-life science studies how chemistry may have become biology, but unanswered questions are not evidence for God.

The human body is remarkable, but it is not perfectly designed.

Disease, suffering, extinction, and biological waste are serious problems for the idea of perfect divine design.

Human beings are not separate from nature. We are part of the living world.

Science does not make life meaningless. It makes life more honest, connected, and precious.

Key Takeaway

Life is beautiful, but beauty is not proof of God. When we examine living beings closely, we do not find the clean signature of perfect divine engineering. We find evolution, inheritance, mutation, adaptation, disease, extinction, and deep time. The living world does not show scientific evidence of a designer. It shows the power of nature.

Sources

National Academies of Sciences, Engineering, and Medicine. Science, Evolution, and Creationism.

University of California Museum of Paleontology. Understanding Evolution.

University of California Museum of Paleontology. Understanding Science.

NASA Astrobiology. Origin-of-Life and Prebiotic Chemistry Resources.

National Human Genome Research Institute. Mutation.

Charles Darwin. On the Origin of Species.

Jerry A. Coyne. Why Evolution Is True.

Richard Dawkins. The Blind Watchmaker.

Neil Shubin. Your Inner Fish.

Transition to the Next Chapter

Before we discuss Darwin, DNA, fossils, disease, and human evolution, we must first ask a basic question:

What would real evidence of a designer actually look like?

Many people say life looks designed. But science cannot stop at appearance. It must ask whether the evidence truly supports the claim.

That is where we begin.

Chapter 1: What Would Evidence of a Designer Look Like?

The Stone in the Desert

Imagine an archaeologist walking across a dry desert plain.

The afternoon sun is sharp. The ground is scattered with ordinary stones, shaped by wind, heat, pressure, and time. Most of them mean nothing. They are only broken rock.

Then she notices one stone that looks different.

It fits the hand. One edge is sharper than the others. The surface carries small repeated marks. The shape is not perfect, but it looks intentional. It may have been struck again and again by another stone. It may have been used to cut meat, scrape hide, or shape wood.

At that moment, the archaeologist faces a serious question.

Was this stone shaped by nature, or was it shaped by a mind?

She does not answer by emotion. She does not say, “It feels designed to me, so it must be designed.” She looks for evidence.

Are there other shaped stones nearby?

Are there marks of repeated tool-making?

Are there bones, ashes, campsites, or signs of human activity?

Does the stone resemble known tools from the same period?

Can natural forces explain its shape, or does human work explain it better?

This is how evidence works.

The question is not only, “Does it look designed?”

The better question is, “What should we expect to find if it was designed, and what do we actually find?”

This chapter begins with that same question, but on a much larger scale.

Many people look at living beings and say, “Life must have been designed by God.”

The eye seems made for seeing.

The wing seems made for flying.

The heart seems made for pumping blood.

The flower seems made for the bee.

The human body seems made for life.

That reaction is understandable. Life can fill the mind with wonder. But wonder is not proof. A feeling of design is not the same as scientific evidence for a designer.

Before we can honestly say that life points to God, we must ask a more disciplined question:

What would real scientific evidence of a divine designer look like?

Once we ask that question carefully, the design argument becomes much weaker.

The First Mistake: Confusing Feeling With Evidence

Human beings often begin with feeling.

We feel that the sun moves across the sky. For thousands of years, that felt obvious. But astronomy showed that the Earth rotates and moves around the sun.

We feel that solid ground is still. But geology shows that continents move.

We feel that disease must have some moral or spiritual meaning. For much of history, people blamed illness on curses, demons, sin, bad air, or divine punishment. But medicine revealed microbes, genetics, cancer, inflammation, immune disease, toxins, and many other natural causes.

The feeling was powerful.

The feeling was wrong.

This matters because the argument from design often begins in the same way. A person looks at life and feels that it must be designed. The feeling may be sincere. It may be emotionally strong. It may come from childhood, family, prayer, culture, and tradition.

But science cannot stop at feeling.

Science asks for evidence. It asks whether an idea can be tested. It asks whether an explanation helps us understand the facts better than other explanations. It asks what we should expect to find if the idea is true.

When a claim concerns the natural world, it must face the evidence of the natural world.

If someone says, "Life looks designed to me," that is a personal impression.

If someone says, "Life provides scientific evidence for God," that is a much stronger claim.

This book examines that stronger claim.

What Science Requires

Science is not a collection of opinions. It is a disciplined way of testing ideas against reality.

A scientific explanation must do more than comfort us. It must connect evidence with cause. It must allow questions. It must be open to correction. It must risk being wrong.

Consider the germ theory of disease.

If germs cause certain diseases, then we should be able to find those germs in infected people. We should be able to study how they spread. We should be able to prevent some infections with hygiene, vaccines, and public health measures. We should be able to treat some infections with antibiotics, antivirals, or other medical tools.

That is exactly what happened.

The germ theory became powerful because it explained, predicted, and helped us act.

Now consider genetics.

If DNA carries inherited information, then children should share more DNA with their biological parents than with strangers. Relatives should share recognizable genetic patterns. Some diseases should follow patterns of inheritance. Genetic testing should reveal relationships.

That is what we find.

Now consider evolution.

If living beings share common ancestry, then they should not appear as unrelated creations. We should find patterns of relationship. Closely related species should share more anatomy and DNA than distant species. Fossils should appear in meaningful historical order. We should find extinct forms. We should see signs of modification, not sudden perfection.

That is also what we find.

Now apply the same standard to divine design.

If life was designed by a perfect, all-knowing, all-powerful, loving creator, what should we reasonably expect to see?

This is a fair question.

It does not mock religion. It simply asks the design claim to meet the same standard that every scientific claim must face.

What Perfect Design Might Lead Us to Expect

If a perfect divine designer created life from nothing, we might expect the living world to show a very different pattern from the one we observe.

We might expect bodies that are safer and more efficient.

We might expect fewer inherited defects.

We might expect no childhood cancers.

We might expect no genetic diseases that slowly destroy a child's muscles, lungs, blood, brain, or metabolism.

We might expect no immune system that sometimes attacks the body it is supposed to protect.

We might expect no dangerous childbirth.

We might expect no blind spot in the eye.

We might expect no wisdom teeth that often do not fit the jaw.

We might expect no appendix that can become inflamed and rupture.

We might expect no parasites that survive by feeding inside other living bodies.

We might expect no world in which one animal must tear another apart in order to live.

We might expect no mass extinction of most species that ever existed.

A believer may answer, "God may have reasons we do not understand."

That may be possible as a statement of faith. But it is not a scientific explanation.

If the claim is that life clearly shows divine design, then the whole living world must be considered, not only the beautiful parts. We cannot count the eye but ignore blindness. We cannot count birth but ignore birth defects. We cannot count healing but ignore cancer. We cannot count flowers but ignore parasites. We cannot count the living species and ignore the vast graveyard of extinct life.

A serious argument must include all the evidence.

What We Actually Find

When we look closely at life, we do not find perfect engineering.

We find working systems full of compromise.

The human body is remarkable, but it is fragile. The spine allows upright walking, but it often causes pain. The throat allows speech, but it also creates a risk of choking. The eye sees beautifully, but it has a blind spot. The reproductive system creates life, but pregnancy and childbirth can be dangerous. The immune system protects us, but it can also create autoimmune disease. Cell division allows growth and repair, but it can become cancer.

This is not what we would normally call flawless design.

It is what we would expect from evolution.

Evolution does not start with a clean blueprint. It modifies what already exists. It does not build the best possible body. It preserves traits that work well enough for survival and reproduction in a particular environment.

That phrase, "well enough," is important.

Evolution does not produce perfection.

It produces survival.

A body can be good enough to reproduce and still be full of future problems. A trait can help early life and still cause trouble later. A system can protect us most of the time and still fail badly in some people. A structure can be useful and still be awkward, vulnerable, or dangerous.

This is the pattern we find in biology.

It looks less like divine engineering and more like natural history.

The Problem of Explaining Everything

A weak explanation can sometimes survive by explaining everything.

This is one of the problems with divine design.

If a child is born healthy, some people call it God's design.

If a child is born with a devastating disease, they say God's plan is mysterious.

If the eye works well, it is evidence of design.

If the eye has a blind spot, that too is somehow part of design.

If a species survives, God preserved it.

If a species goes extinct, God had a reason.

If nature is beautiful, it shows God's love.

If nature is cruel, we are told not to question God.

This may protect belief emotionally, but it weakens the claim scientifically.

A scientific explanation must be able to face possible failure. It must tell us what would count against it. If every possible result is called design, then design explains nothing in a scientific sense.

It becomes a label placed on whatever happens.

Evolution is different.

Evolution makes real expectations. If humans were found in the oldest fossil layers, that would be a serious problem for evolution. If DNA showed no pattern of relationship among species, that would be a serious problem. If living beings did not change at all, if fossils appeared in random order, or if species were completely unrelated in anatomy and genetics, evolution would be in trouble.

But that is not what we find.

We find the broad pattern evolution predicts.

Design and Natural Order Are Not the Same

Some people think that order in nature automatically means design.

But order and design are not the same.

Snowflakes have order, but they are not carved by angels. Crystals have structure, but they are not built by a conscious mind. A river can carve a canyon without knowing what it is doing. Gravity can gather gas into stars without intention. Physical laws can create patterns.

Natural processes can produce order.

Biology adds another powerful process: natural selection.

Natural selection can preserve useful inherited traits over generations. A feature that helps survival or reproduction may become more common. Over deep time, this can produce organs, instincts, camouflage, teeth, wings, roots, and countless other adaptations.

There is no mind behind the process.

There is no foresight.

There is no mercy.

But there is filtering.

This is why life can look purposeful without being planned.

A bird's wing looks made for flying because earlier structures were modified across generations in ways that helped movement, balance, gliding, display, insulation, or flight. An eye looks made for seeing because light-sensitive structures gave advantages and were gradually refined. A flower looks made for a bee because plants and pollinators influenced one another over long periods.

The final result can look designed.

The process does not require a designer.

Why the Watchmaker Comparison Fails

The old design argument often uses human-made objects.

A watch has a watchmaker.

A house has a builder.

A computer has a programmer.

Therefore, life must have a designer.

This sounds simple, but it is a poor comparison.

A watch does not grow.

A house does not reproduce.

A computer does not pass inherited variation to offspring.

A machine does not form a population that changes over generations through natural selection.

Living beings are different.

They reproduce. They inherit. They vary. They mutate. They compete. They adapt. They die. Their descendants carry modified versions of what came before.

A watch is assembled from outside.

A living organism develops from within.

If watches produced slightly different watches, and if some of those watches survived better in certain environments and produced more copies, then the comparison would be closer. But watches do not do that.

Living beings do.

This is why the watchmaker argument fails. It treats organisms as if they were manufactured objects. They are not. They are historical systems.

They carry ancestry.

Why Imperfection Matters

Some religious defenders say that imperfections in nature do not disprove God.

That is true in a limited sense. One imperfect structure does not disprove every possible God. But that is not the strongest argument.

The stronger argument is this:

The pattern of life looks exactly like what evolution would produce, and it does not look like what perfect divine engineering would lead us to expect.

This is not about one blind spot, one painful spine, one bad nerve route, one disease, or one extinct species. It is about the whole pattern.

Life is useful but not perfect.

Bodies are functional but vulnerable.

Genes carry information but also errors.

Organs work but also fail.

Species adapt but also disappear.

Nature contains beauty but also immense suffering.

Humans are intelligent but biologically related to other animals.

This whole pattern fits evolution.

It fits natural history, not perfect divine design.

If a believer says every flaw is part of a mysterious plan, the claim moves outside science. It may remain faith, but it is no longer evidence.

God of the Gaps Is a Weak Foundation

Religion has often placed God where knowledge was missing.

When people did not understand thunder, gods were placed in storms.

When people did not understand disease, spirits and punishments were imagined.

When people did not understand the planets, divine movement was invoked.

When people did not understand species, special creation was assumed.

Again and again, science replaced supernatural guesses with natural explanations.

This pattern should make us careful.

A gap in knowledge is not evidence for God.

A mystery is not a miracle.

The honest answer to an unanswered question is not "therefore God." The honest answer is, "We do not yet know, so we continue to investigate."

This is the difference between science and faith.

Science keeps the question open.

Faith often closes it too soon.

The claim "God designed it" may feel emotionally satisfying, but it usually stops the search. It does not tell us how the body developed, how DNA changed, how fossils formed, how disease arises, or why species go extinct.

Science asks those questions.

That is why science grows.

If God Leaves No Evidence

Some believers respond by saying God works invisibly behind natural processes.

God designed through evolution.

God guided mutation.

God arranged natural selection.

God created the laws.

God acted in ways science cannot detect.

But if God's action cannot be detected, measured, tested, or distinguished from no action at all, then it is not a scientific explanation.

A person may believe it privately. That is faith.

But it cannot be presented as scientific evidence.

This distinction is important.

Science does not say that every possible hidden God has been disproven in every philosophical sense. Science says that such a God is unnecessary as an explanation of life.

Evolution works without adding God. Genetics works without adding God. Medicine works without adding God. Paleontology works without adding God. Comparative anatomy works without adding God. The tree of life works without adding God.

If God makes no testable difference, then God has left the field of science.

That is not anger.

It is clarity.

A Better Explanation

Evolution explains the living world better than divine design because it explains both success and failure.

It explains why organisms are adapted to their environments.

It explains why adaptation is not perfection.

It explains why bodies contain awkward structures.

It explains why genes show shared ancestry.

It explains why fossils appear in historical order.

It explains why species change.

It explains why extinction is common.

It explains why humans are part of the primate family.

It explains why disease is built into biological life.

Divine design does not explain these facts better. Often, it explains them less clearly.

If humans were specially created, why do we fit so well within the primate branch of life?

If species were separately created, why do they form nested patterns of relationship?

If DNA is a perfect divine code, why does it contain broken genes, mutations, shared errors, and ancient viral traces?

If nature reflects divine love, why is so much of life built on suffering, predation, parasitism, and death?

These questions are not disrespectful.

They are necessary.

A belief about the natural world must answer to the evidence of the natural world.

The Calm Conclusion

The purpose of this chapter has been simple.

Before we ask whether life was designed, we must ask what would count as evidence of design.

When we do that honestly, the design argument weakens.

Life can feel designed, but feeling is not evidence. Life can be beautiful, but beauty is not proof. Life can be complex, but complexity is not automatically divine. Life can show order, but natural processes can create order.

When we look closely, life does not show the clean signature of perfect engineering.

It shows adaptation, ancestry, mutation, compromise, disease, extinction, and change.

This is exactly what evolution would lead us to expect.

It is not what perfect divine design would lead us to expect.

The conclusion is clear:

Life does not provide scientific evidence for God.

Life provides evidence for natural processes.

No designer is required.

Chapter Summary

A feeling of design is not the same as scientific evidence.

Science asks what we would expect to observe if a claim were true, then compares that expectation with reality.

If life were designed by a perfect, loving, all-powerful creator, we might expect fewer harmful flaws, less biological suffering, fewer genetic errors, and less waste.

What we actually find is a living world filled with adaptation, but also disease, genetic error, predation, extinction, and imperfect bodies.

Natural order is not the same as divine design.

The watchmaker comparison fails because living beings reproduce, vary, inherit, and evolve, while watches do not.

The design argument often explains everything after the fact, which makes it weak as science.

Evolution explains both the beauty and the imperfection of life better than divine design.

A God who leaves no testable evidence is not a scientific explanation.

Key Takeaway

The first step in examining divine design is to ask what real evidence of a designer would look like. When we ask that honestly, life does not show the clean signature of perfect engineering. It shows natural history: adaptation, inheritance, compromise, suffering, extinction, and change. The living world does not give us scientific evidence for God. It gives us evidence for evolution.

Sources

National Academies of Sciences, Engineering, and Medicine. Science, Evolution, and Creationism.

National Academies Press. Evidence Supporting Biological Evolution.

University of California Museum of Paleontology. Understanding Science.

University of California Museum of Paleontology. Understanding Evolution.

National Center for Science Education. Evolution, Creationism, and Intelligent Design Resources.

Charles Darwin. On the Origin of Species.

Richard Dawkins. The Blind Watchmaker.

Jerry A. Coyne. Why Evolution Is True.

Kenneth R. Miller. Finding Darwin's God.

Transition to the Next Chapter

Now that we have asked what real evidence of design would require, we can look more closely at why the design argument feels so powerful.

Why does life look designed?

Why does the eye seem made for seeing, the wing for flying, the flower for attracting insects, and the body for survival?

The answer is not God.

The answer is one of the greatest ideas in science:

Natural selection.

Chapter 2: Why Life Looks Designed

The Eye That Misleads Us

Look into the eye of another human being.

It is difficult not to feel wonder.

The clear cornea bends light. The pupil widens and narrows. The lens focuses. The retina receives color, shape, brightness, and movement. Nerves carry signals to the brain. In a fraction of a second, the world appears: a face, a written word, a red light, a child's smile, the soft gold of evening, the danger of an approaching car.

To the ordinary mind, the eye seems made for seeing.

That is where the old religious argument begins.

If there is an eye, there must be an eye-maker.

If there is a wing, there must be a wing-maker.

If there is a heart, there must be a heart-maker.

If life fits the world so beautifully, someone must have fitted it.

This thought feels natural. It feels simple. It feels emotionally powerful.

But it is not scientifically necessary.

The eye was not created in one supernatural act. Wings were not drawn on a heavenly blueprint. The heart was not carved like a pump by a divine engineer. Living organs were not assembled from outside like machines.

They developed through the long natural history of life.

This is one of the most important ideas in this book:

Life looks designed because natural selection can create the appearance of design without a designer.

That sentence changes everything.

It allows us to admire the eye without worshiping an eye-maker. It allows us to study the wing without imagining a wing-maker. It allows us to respect the body without turning biology into theology.

Life may look designed.

But looking designed is not the same as being designed.

Why the Design Argument Feels Strong

The design argument has survived for centuries because it begins with something real.

Life is astonishing.

A bird migrates across continents. A bee finds flowers and returns to the hive. A seed lies silent in the soil and then rises toward light. A wound closes. A fever fights infection. A fetus develops from one fertilized cell into a child with eyes, lungs, fingers, blood vessels, and a brain.

No honest person should pretend that these things are not remarkable.

The mistake is not wonder.

The mistake is jumping from wonder to God.

Religion often makes that jump quickly. It says, "This is beautiful, therefore God." Or, "This is complex, therefore God." Or, "This is useful, therefore God designed it."

But usefulness is not proof of divine design.

A river may be useful to animals, trees, and human beings, but the river was not dug by a god. A cave may shelter an animal, but it was not built for that animal. The moonlight may help a traveler at night, but the moon was not placed there for that traveler.

Natural things can have useful effects without being intentionally arranged.

Living beings are more complex than rivers and caves because they reproduce, inherit traits, vary, and change over generations. That difference makes evolution possible. Traits that help survival or reproduction can become more common. Traits that fail can disappear.

The result may look planned.

But it is not planned.

It is selected.

The Mind Searches for Purpose

There is another reason life looks designed to us.

The human mind is built to see intention.

If a branch cracks in the dark, we wonder who is there. If grass moves suddenly, we imagine an animal. If a shadow passes across a window, we look for a person. This quick search for agency may have helped our ancestors survive.

A person who mistook the wind for a predator lost little.

A person who mistook a predator for the wind might not survive.

So the mind became sensitive to agents, intentions, and hidden causes.

That habit still lives inside us. We see faces in clouds. We see messages in coincidence. We see purpose in accidents. We ask why something happened even when there may be no personal reason behind it.

A child becomes ill, and someone asks, "Why did God do this?"

A storm destroys homes, and someone says, "It must be punishment."

A person survives an accident, and others say, "God saved him."

These reactions come from the human hunger for meaning. They do not come from scientific evidence.

The same habit affects how people look at life.

When we see a flower attracting bees, we imagine intention. When we see camouflage protecting an insect, we imagine planning. When we see a mother bird feeding her chicks, we imagine a loving design in nature.

But nature does not need intention to produce these results.

Flowers and pollinators can influence each other over generations. Predators and prey can shape each other. Parasites and hosts can shape each other. Climate, food, disease, competition, and mating can all filter inherited traits.

Over time, life becomes fitted to life, and life becomes fitted to its surroundings.

From a distance, it looks like design.

From close examination, it looks like evolution.

"Made For" Does Not Mean "Made By"

Our language can mislead us.

We say the eye is made for seeing.

We say wings are made for flying.

We say teeth are made for chewing.

We say the heart is made for pumping blood.

This language is convenient, but it quietly suggests intention.

When we say "made for," we may begin to imagine "made by."

But in biology, "made for" usually means something different. It means a structure has a function because it was shaped, preserved, or modified through natural history.

A bird's wing is for flying in the sense that flight is one of its functions now. But that does not mean a divine engineer designed it in advance. Wings developed from earlier structures. Feathers may have served other purposes before they became central to powered flight. Forelimbs changed over time. Small advantages accumulated.

The final result became flight.

The same is true for many organs and behaviors.

A structure can have a function without having a conscious maker.

We understand this in ordinary life. A rock under a tree may become a seat. That does not mean someone designed the rock as a chair. A fallen log may become a bridge across a stream. That does not mean it was built as a bridge. A cave may become a shelter. That does not mean it was made for shelter.

In biology, the process is more powerful because inherited variation and natural selection can refine living structures across generations.

A light-sensitive patch of cells can help an organism detect light and dark. A shallow cup of light-sensitive cells can help detect direction. A deeper cup can improve the information. A transparent covering can protect delicate cells. A lens-like structure can improve focus.

Step by step, better vision can evolve.

The eye did not need to appear fully formed.

Useful intermediate stages are enough.

That is why the old question, "What use is half an eye?" misses the point. A simple light-sensitive system can be much better than no vision at all. In evolution, partial usefulness matters.

Adaptation: The Source of the Illusion

Adaptation is one reason the living world feels designed.

A fish is shaped for water.

A bird is shaped for air.

A cactus is shaped for dryness.

A polar bear is shaped for cold.

A bat is shaped for night.

A hummingbird is shaped for flowers.

A cheetah is shaped for speed.

A spider is shaped for web-making.

This fit between organism and environment can look purposeful. It can feel as if each living thing was assigned its role.

But adaptation does not require assignment.

It requires variation, inheritance, selection, and time.

In a dry environment, plants that conserve water better are more likely to survive and reproduce. Over many generations, features that reduce water loss can become more common. Thick stems, reduced leaves, waxy surfaces, and deep roots do not appear because the desert planned them. They persist because they helped survival in that environment.

In cold climates, animals with better insulation may survive better. In places with predators, camouflage may help prey avoid death. In places where flowers and insects interact, both can change over time in ways that influence each other.

This is the quiet power of evolution.

It does not create perfection.

It creates fit.

And that fit is local. A trait that helps in one environment may be useless or harmful in another. Thick fur helps in the Arctic but can be dangerous in heat. A long beak may help with one flower but not another. A large body may help against some predators but require more food.

A perfect designer would not need to work through such fragile local compromises.

Evolution does.

Camouflage and the Illusion of Planning

Few things look more designed than camouflage.

A leaf insect looks like a leaf.

A stick insect looks like a twig.

A snowshoe hare turns white in winter.

A flounder blends into the sea floor.

A moth resting on bark can almost disappear.

It is tempting to say, "Surely someone planned this."

But camouflage can evolve through small differences filtered by survival.

Imagine a population of insects living among leaves. Some are slightly greener. Some are slightly more leaf-shaped. Birds eat the insects they can see more easily.

The insects that are harder to see survive a little more often and leave more offspring. Their offspring inherit some of those traits. Over many generations, the population becomes more leaf-like.

No conscious painter is needed.

Predation itself becomes the filter.

Nature does not say, "Let me design a leaf insect." Nature simply removes more of the insects that are easy to see. The ones less easily seen survive more often. Over time, survival changes the population.

The result may look artistic.

The process is not artistic.

It is selection.

Religious arguments often focus on the final result and ignore the process. They point to the leaf insect and say, "Look how perfect." Evolution asks a better question:

What repeated pressures could shape such a form over time?

That question leads to evidence.

Instinct Without a Divine Script

Animal behavior can also look designed.

Birds build nests without attending architecture school. Sea turtles hatch on beaches and move toward the ocean.

Salmon return to spawning grounds. Bees communicate the location of food. Spiders spin webs. Ants build colonies. Monarch butterflies migrate across great distances.

These behaviors can make nature feel guided by hidden intelligence.

But instinct does not require a divine script.

Behavior can be shaped by evolution. If certain inherited tendencies help survival or reproduction, they can become more common. An animal that responds to light, smell, temperature, sound, or seasonal change in a useful way may survive better than one that does not.

Animal behavior is complex. Genes, development, learning, environment, and social interaction can all matter. But none of this requires God as an explanation.

A bird does not need to understand engineering to build a nest. A spider does not need to know geometry to spin a web. Natural selection can preserve patterns of behavior that work.

There is beauty here, but not magic.

There is complexity, but not divine command.

The Flower and the Bee

Consider a flower and a bee.

The flower has color, scent, shape, nectar, and pollen. The bee has eyes, wings, legs, mouthparts, and behavior that bring it to the flower. The bee gets food. The flower gets pollinated. The relationship looks arranged.

For many people, this feels like evidence of God.

But biology gives a stronger explanation.

Flowers that attracted pollinators more successfully left more descendants. Insects that found food more efficiently survived and reproduced. Over time, plants and pollinators influenced each other.

The flower did not need to know the bee.

The bee did not need to understand the flower's reproductive system.

No invisible hand needed to adjust each detail.

The relationship was shaped by mutual pressure across generations.

This is called coevolution.

The same broad principle appears in predator and prey, parasite and host, virus and immune system, plant and pollinator. Life is constantly shaping life.

This does not look like a static creation placed on Earth fully formed.

It looks like history.

The Human Body: Impressive, But Not Perfect

The human body is often used as evidence of design.

And it is easy to understand why.

The heart pumps. The lungs exchange gases. The kidneys filter blood. The liver performs complex chemical work. The skin protects. The immune system fights infection. The brain receives, organizes, and responds to the world. The reproductive system allows life to continue.

As a physician, I have seen the body's power many times. I have seen lungs recover, wounds close, infections clear, blood pressure improve, hearts stabilize, kidneys resume function, and patients return from the edge of death.

The body deserves respect.

But respect is not worship.

The body is remarkable, but it is not perfect. It is full of trade-offs and vulnerabilities. The same immune system that saves us can attack our joints, thyroid, intestines, skin, nerves, or kidneys. The same cell division that allows a child to grow can later produce cancer. The same clotting system that prevents bleeding can form a clot that blocks a coronary artery or travels to the lungs. The same spine that allows upright walking can produce lifelong pain.

A perfect engineer could avoid many of these problems.

Evolution cannot always avoid them because evolution works with what already exists. It modifies earlier systems. It favors traits that help survival and reproduction, not necessarily comfort, fairness, or long life.

The body looks designed if we look only at its successes.

It looks evolved when we include its failures.

Adaptation Is Not Perfection

This distinction is essential:

Evolution produces adaptation, not perfection.

An adaptation is a feature that helps survival or reproduction in a particular environment. It does not have to be ideal. It only has to work well enough.

This explains much of biology.

The human knee works, but it can fail.

The spine supports us, but it can ache.

The eye sees, but it has a blind spot.

The immune system protects, but it can overreact.

The reproductive system continues the species, but it can be dangerous and painful.

The brain is powerful, but it can suffer anxiety, depression, addiction, delusion, and dementia.

If we expect perfect design, these facts are troubling.

If we understand evolution, they are not surprising.

Evolution has no duty to create comfort. It has no moral concern. It does not care whether an organism suffers. It does not care whether an old person has back pain after reproduction is complete. It does not care whether a child is born with a genetic disease. It does not care whether a gazelle dies in terror in the mouth of a lion.

Evolution is not cruel in the human sense, because cruelty requires intention.

It is simply indifferent.

That is why life can be wonderful and terrible at the same time.

Why "God Designed It That Way" Explains Nothing

Some people respond to every biological feature by saying, "God designed it that way."

But this does not explain how anything works.

If a flower attracts a bee, "God designed it" does not explain pollination, genetics, selection, plant reproduction, or coevolution.

If the eye sees, "God designed it" does not explain development, anatomy, optics, nerve signals, brain processing, or evolutionary history.

If the immune system fights infection, "God designed it" does not explain antibodies, white blood cells, inflammation, immune memory, or autoimmune disease.

"God designed it" is not an explanation in biology.

It is a stopping point.

Science does not stop. It asks how, by what mechanism, under what conditions, with what evidence, and with what predictions.

This is why evolution is so powerful. It does not merely say that life changed. It explains mechanisms by which life changes. It connects fossils, anatomy, DNA, development, geography, observed evolution, and medicine.

The design argument gives a feeling of explanation.

Evolution gives a working explanation.

We Usually See Only the Survivors

There is another reason the illusion of design is so strong.

We usually see the survivors.

We see the birds that fly, not all the failed branches that disappeared. We see the animals alive today, not the countless extinct species buried in rock. We see the successful pregnancy, not every miscarriage. We see the healthy child, not every genetic error that prevented life from continuing. We see the flower in bloom, not every seed that failed to grow.

Nature is full of filtering.

The living world around us is the result of immense selection. What failed often vanished. What survived remains visible. So the world can appear more polished than it really is.

It is like walking through a museum that displays only the winning designs and hides the graveyard of failures.

But fossils bring back some of the vanished world. Genetics reveals old errors and inherited traces. Disease exposes the body's fragility. Extinction reminds us that life's history is not a smooth plan but a long record of survival and disappearance.

Once we include the failures, the illusion of design weakens.

Life no longer looks like perfect creation.

It looks like a harsh and beautiful process in which some forms survived and many did not.

Explanation Is Better Than Comfort

Religion often offers comfort.

Science offers explanation.

These are not the same.

When a child asks, “Who made the birds?” the religious answer may feel warm and simple: “God made them.” But that answer closes the door too soon. It does not explain feathers, bones, flight muscles, eggs, migration, fossils, DNA, or the relationship between birds and dinosaurs.

The scientific answer is more demanding, but also more honest.

Birds are living descendants of earlier forms of life. Their bodies carry traces of ancestry. Their wings developed from forelimbs. Their feathers had evolutionary histories. Their bones, lungs, eggs, and genes connect them to the larger tree of life.

This explanation does not make birds less beautiful.

It makes them more amazing.

A bird is no longer a finished object placed into the world. It is a living chapter in a story billions of years old.

That story does not require God.

It requires nature, time, inheritance, variation, selection, and survival.

A Mature Kind of Wonder

Some people fear that without design, life becomes cold.

That fear is understandable, but unnecessary.

A scientific view does not turn the living world into dead machinery. It reveals the depth of life's history. It allows us to see that every organism carries a past. The flower is not just a decoration. It is part of a long story of plants, pollinators, sunlight, soil, and survival. The bird is not just an object in the sky. It carries the history of dinosaurs, feathers, flight, and extinction. The human body is not a clay figure shaped by divine hands. It is a living record of ancestry.

This is not less wonderful.

It is more wonderful.

But it is wonder without worship.

It is wonder that does not stop at God.

It continues toward understanding.

The Central Point

Life looks designed because natural selection can create useful complexity over time.

This is not a small idea. It is one of the great turning points in human thought.

Before evolution, the design argument seemed strong because there was no known natural process powerful enough to explain adaptation. The eye seemed impossible. The wing seemed impossible. The flower and the bee seemed impossible. The human body seemed impossible.

After evolution, the situation changed.

Now we have a natural explanation.

This does not mean every religious person immediately stops believing. Belief is emotional, cultural, personal, and often deeply rooted. But intellectually, the design argument loses its foundation. Once natural selection explains the appearance of design, God is no longer needed as the explanation.

The living world remains beautiful.

But it is not scientific evidence for God.

Chapter Summary

Life looks designed because living beings often fit their environments in impressive ways.

The human mind naturally sees purpose and intention, even where natural processes are enough.

"Made for" does not necessarily mean "made by."

Adaptation can arise through variation, inheritance, natural selection, and time.

Camouflage, instincts, pollination, and complex organs can look planned, but they can develop through evolutionary processes.

The human body is remarkable, but it is also imperfect, vulnerable, and full of trade-offs.

Evolution produces adaptation, not perfection.

The design argument often gives emotional comfort, but it does not provide a testable biological explanation.

Natural selection explains the appearance of design without requiring God.

Key Takeaway

Life looks designed because evolution can shape useful features over long periods of time. The eye, the wing, the flower, the instinct, and the human body may appear planned when viewed from a distance. But when we look closely, we find natural selection, inheritance, variation, compromise, and history. The appearance of design is real. A divine designer is not required.

Sources

Charles Darwin. On the Origin of Species.

National Academies of Sciences, Engineering, and Medicine. Science, Evolution, and Creationism.

University of California Museum of Paleontology. Understanding Evolution: Natural Selection.

University of California Museum of Paleontology. Understanding Evolution: Adaptation.

University of California Museum of Paleontology. Understanding Evolution: Looking at Complexity.

Dan-Eric Nilsson. Eye Evolution and Its Functional Basis.

Richard Dawkins. The Blind Watchmaker.

Jerry A. Coyne. Why Evolution Is True.

Neil Shubin. Your Inner Fish.

Sean B. Carroll. Endless Forms Most Beautiful.

Transition to the Next Chapter

If life looks designed today, it is easy to understand why people in the past believed in creation.

Before fossils were properly understood, before deep time was accepted, before DNA was discovered, and before natural selection was known, the living world seemed to demand a supernatural explanation.

The next chapter will take us into that earlier world, when life still seemed to need a creator.

Chapter 3: Before Darwin, Life Needed a Creator

A World Full of Living Mysteries

Imagine standing in a field three hundred years ago.

There is no knowledge of DNA. No understanding of genes. No modern fossil record. No radiometric dating. No clear idea that the Earth is billions of years old. No evolutionary tree. No explanation for how species change through natural processes.

You look around and see life everywhere.

Birds cross the sky. Bees move from flower to flower. Trees rise from soil. Cattle graze in the distance. A seed becomes a plant. A caterpillar becomes a butterfly. A child is born. An old man dies. A wound heals. A fever kills.

Life appears, grows, reproduces, suffers, and disappears.

What explanation would feel most natural?

For most of human history, the answer was simple:

Someone made all this.

That answer was not foolish for its time. It was human. Ancient people had deep questions but limited tools.

They saw beauty, birth, death, disease, order, and the astonishing variety of living things. They wanted meaning. They wanted comfort. They wanted an explanation that made the world feel less mysterious.

So they created stories.

Gods shaped animals.

Gods formed human beings.

Gods placed plants upon the earth.

Gods gave each creature its purpose.

Gods explained why humans seemed different from other living beings.

Creation stories became the first biology of ancient humanity.

But they were not science.

They were early human attempts to explain a world that human beings did not yet have the tools to understand.

When Creation Felt Obvious

Before Darwin, life seemed to require a creator because no strong natural explanation existed.

A horse looked like a horse. A lion looked like a lion. A fish looked like a fish. A human being looked separate from all of them. Species appeared stable, complete, and distinct.

Dogs gave birth to puppies. Cats gave birth to kittens. Birds hatched birds. Human beings gave birth to human beings. In one lifetime, people did not see one kind of organism slowly becoming another. They saw resemblance, but they also saw continuity.

So the idea of fixed species felt reasonable.

Religious traditions strengthened that impression. They taught that living beings were created in separate kinds, each with its own place. Human beings were often placed above animals. Animals were placed below humans. Nature was seen as a designed order, and God was seen as the maker of that order.

This gave people certainty.

It also gave religion power.

If God created life, then religious authorities could claim to speak not only about worship, but also about human identity, morality, nature, social order, gender roles, and the meaning of life itself.

To question creation was not merely to question biology.

It was to question the whole religious picture of reality.

That is why evolution later became so disturbing. It did not challenge one small belief. It challenged the foundation of an entire worldview.

The Comfort of a Fixed World

The old view of creation had emotional strength.

It told people that the world was not accidental. It told them that every creature had a place. It told them that human beings were special. It told them that suffering, death, and mystery were part of a larger divine plan.

This is important to understand.

Many people do not believe in creation because they have carefully compared scientific evidence. They believe because the story gives comfort. It tells them they were intended. It tells them someone made them. It tells them life has a purpose beyond nature. It tells them humanity is not merely another animal species.

That emotional power is real.

But emotional power is not the same as truth.

A story can comfort us and still be wrong. A belief can be ancient and still lack evidence. A tradition can be sacred and still fail as science.

The fixed world of creation gave simple answers to difficult questions.

Where did life come from? God made it.

Why are humans special? God created them in His image.

Why do creatures fit their surroundings? God designed them.

Why does nature seem ordered? God arranged it.

Why do we suffer? God has a reason.

These answers felt complete, but they did not explain mechanisms. They did not predict fossils. They did not reveal genes. They did not explain extinction. They did not show why humans share anatomy and DNA with other animals.

They closed the question too soon.

Science would reopen it.

The Ladder That Put Humans Near the Top

Long before modern biology, many cultures imagined life as a hierarchy.

At the top was God or the gods. Below them were spiritual beings. Then came human beings. Below humans were animals. Below animals were plants. Below plants were stones, soil, and the nonliving world.

In Western thought, this idea was often called the Great Chain of Being.

It was not evolution.

It was a ladder.

This ladder placed humans near the top of reality. It separated us from the rest of nature and made us feel spiritually superior. It gave the world a sense of order, but it also encouraged arrogance. If humans were specially created above animals, then animals could be treated as lower beings. If the Earth was made for humans, nature could be treated as a possession.

Evolution later replaced the ladder with a tree.

That change was revolutionary.

Life is not a divine staircase with humans near the top. Life is a branching tree with many limbs, many dead branches, and no single species as the purpose of the whole structure.

Humans are unusual. We have language, science, memory, art, morality, and culture. But we are not separate from nature. We are one branch of life.

This shift from ladder to tree is one of the great turning points in human thought.

It also weakens the religious idea of special creation.

Naming Life Before Explaining Life

Before Darwin, naturalists did valuable work.

They collected plants, insects, shells, bones, birds, and animals. They described living beings carefully. They compared forms. They created systems of classification. They noticed that life was not random.

Cats resembled other cats. Horses resembled other horses. Birds shared bird-like features. Mammals shared mammalian features. Apes resembled humans in ways many people found uncomfortable.

Classification helped organize life.

But at first, it did not explain life.

Many people believed classification revealed the order in God's mind. God had created different categories, and naturalists were simply discovering those sacred categories.

This was an important step, but it was incomplete.

Naming is not the same as explaining.

To name a species is not to explain where it came from. To place animals into groups is not to prove that God made those groups separately. To arrange life into categories is not to show divine design.

Evolution later gave classification a deeper meaning.

Living beings fall into groups because they are related. Mammals share features because they inherited them from mammalian ancestors. Primates share features because they inherited them from primate ancestors. Humans resemble other great apes because we share ancestry with them.

The old creation view could organize life.

Evolution could explain it.

Fossils Begin to Disturb the Story

Fossils created a serious problem for fixed creation.

At first, fossils were misunderstood. Some people thought they were strange stones. Some thought they were remains from a great flood. Some tried to force them into familiar religious stories. But as more fossils were found, the problem became harder to ignore.

Many fossils did not resemble any living animal.

There were giant bones of creatures no one had seen alive. There were shells far from the sea. There were rock layers containing forms of life different from those in other layers. There were traces of vanished worlds.

This raised difficult questions.

If God created all species perfectly, why did so many disappear?

If each creature had a divine purpose, why were entire groups erased?

If life was fixed, why did fossils show different forms in different layers?

If the Earth was young, why did the rocks seem to preserve such an ancient history?

Extinction was especially disturbing.

For a fixed creation view, extinction was hard to explain. Why would a perfect creator make species and then allow them to vanish forever? Why create entire worlds of life only to bury them in stone?

Fossils began to say something that later biology would say more clearly:

Life has a history.

And that history is not simple, fixed, or centered on human beings.

Extinction Shakes the Idea of Perfect Creation

Extinction is not a small detail in the story of life.

It is one of the major facts of life's history.

Many species that once existed are gone. Some vanished quietly. Some disappeared in great catastrophes. Some were complex, beautiful, and successful for millions of years. Then they ended.

This fact is difficult for the idea of perfect divine design.

If a perfect God created each species with care and purpose, why create so many forms of life only to erase most of them? Why allow entire branches of life to flourish and then disappear? Why create a world in which survival depends so heavily on changing environments, accident, and catastrophe?

A religious believer may answer, "God had reasons."

But that is not a scientific explanation.

It does not explain the fossil sequence. It does not explain why certain species appear in certain layers. It does not explain mass extinctions. It does not explain environmental change, asteroid impact, volcanic activity, sea-level change, climate shifts, competition, or ecological collapse.

Natural explanations do.

Extinction makes more sense in an unguided natural world than in a perfectly planned divine one.

A world designed for life should not look so much like a graveyard.

Deep Time Opens the Door

The discovery of deep time changed everything.

If the Earth were only a few thousand years old, as many religious traditions once taught, evolution would seem impossible. There would not be enough time for small inherited changes to accumulate into major biological differences.

But geology revealed an ancient Earth.

Rocks showed layers upon layers. Mountains rose and eroded. Rivers carved valleys. Seas advanced and retreated. Volcanoes built land. Sediments accumulated slowly. Fossils appeared in ordered sequences. The planet was not young. It was ancient beyond ordinary human imagination.

This was devastating for simple creation stories.

It stretched the history of nature far beyond ancient scripture. It showed that the Earth existed through vast ages before human beings appeared. It showed that life had come and gone through long periods of change.

Deep time did not prove evolution by itself.

But it made evolution possible.

It gave nature the one thing gradual change requires:

Time.

A single drop of water seems weak. But drops falling for ages can wear stone. A small stream seems harmless. But across deep time, water can carve a canyon. A small inherited difference may seem trivial in one generation. But across millions of years, such differences can reshape life.

Before deep time, creation seemed necessary.

After deep time, natural history became imaginable.

The Problem of Separate Creation

The idea of separate creation says that God created different kinds of living beings independently.

At first glance, this may seem to fit ordinary experience. Dogs remain dogs. Cats remain cats. Birds remain birds. Human beings remain human beings.

But when living beings are compared carefully, separate creation becomes less convincing.

Why should a human arm, a bat wing, a whale flipper, and a cat leg share the same underlying bone pattern?

Why should embryos of different vertebrates show deep similarities during development?

Why should living beings form nested groups, like branches within branches?

Why should humans share so much anatomy with other primates?

Why should DNA later confirm many of the same relationships suggested by anatomy and fossils?

Separate creation can always say, "God chose to make them that way."

But that answer explains nothing.

Common ancestry explains the pattern more naturally.

Similarities exist because living beings inherited modified versions of older structures from shared ancestors. A bat wing, whale flipper, cat leg, and human arm are different uses of related anatomy. They are not separate inventions from nothing. They are variations on inherited biological material.

This is exactly what evolution predicts.

It is not what separate creation naturally suggests.

Human Specialness Before Science

For many religions, the most important claim was not merely that animals were created.

It was that humans were specially created.

Human beings were placed at the center. We were said to have souls. We were said to be made in the image of God. We were given dominion. We were separated from animals by divine intention.

This idea is emotionally powerful.

It tells us we are not just one species among many. It tells us we have cosmic importance. It tells us that the universe has human beings in mind. It tells us that our lives are watched by heaven.

But emotional comfort is not evidence.

As science advanced, the biological separation between humans and other animals became harder to defend. Human beings share anatomy with other mammals. We share skeletal patterns, organs, blood, nerves, reproductive biology, embryological features, and DNA.

We are not outside life.

We are deeply inside it.

The claim of special creation did not fall in one day. It weakened slowly as evidence accumulated. Comparative anatomy, fossils, embryology, genetics, and molecular biology all pointed in the same direction.

Humans are part of nature.

That truth may be difficult for religious pride, but it is not an insult.

We do not need to be separately created to have value.

Why Creation Stories Persist

If science has made special creation unnecessary, why do creation stories still persist?

They persist because they do more than explain biology.

They give identity. They connect people to family, childhood, culture, ritual, and community. They provide comfort in grief. They tell people that life has a purpose. They place human beings inside a story larger than birth and death.

Science explains how life changes.

Religion tells people why they matter.

Science explains disease through biology.

Religion may tell people that suffering has meaning.

Science explains humans as evolved animals.

Religion tells them they were created for a divine purpose.

For many people, giving up creation feels like giving up dignity, comfort, and moral direction.

That is why the argument must be handled with care.

But care does not mean surrendering truth.

We can understand why creation stories persist and still say clearly that they are not scientific explanations. We can respect the emotional attachment and still reject the claim. We can acknowledge the comfort of the story and still follow the evidence.

The purpose is not to humiliate faith.

It is to show that faith is not evidence.

Before Darwin, God Occupied the Gap

Before Darwin, there was a large gap in human knowledge.

People did not know how complex life could arise naturally. They did not know how species could change. They did not know how adaptation could form without intention. They did not know how humans were connected to other animals.

God occupied that gap.

This is a repeated pattern in history.

Before astronomy, gods moved the heavens.

Before meteorology, gods controlled storms.

Before germ theory, spirits and divine punishment explained disease.

Before neuroscience, demons explained seizures and mental illness.

Before evolution, God explained life's diversity.

Darwin did not answer every question about life. No single scientist does that. But Darwin closed one of the greatest gaps in human understanding. He showed that living beings could be explained through natural processes.

After Darwin, creation was no longer needed as the explanation for the diversity of life.

That is why his idea was so dangerous to religious authority.

It did not merely add a new fact.

It removed a divine job.

Origin Stories and Evidence Are Not the Same

Every culture tells origin stories.

These stories may be poetic, meaningful, and historically important. They may show how a people understood themselves. They may express fear, gratitude, longing, hope, and moral lessons. They may bind communities together.

But origin stories are not scientific evidence.

A creation story does not become scientifically true because it is ancient. It does not become factual because millions believe it. It does not become biology because it is sacred. It does not become evidence because it comforts.

Scientific evidence must be testable, observable, and open to correction.

Fossils can be studied.

DNA can be compared.

Anatomy can be examined.

Embryology can be observed.

Genes can be sequenced.

Evolution can be measured.

Predictions can be tested.

Creation stories cannot compete at that level.

They belong to culture, literature, religion, and psychology.

They do not belong to scientific explanation.

A More Mature Wonder

Some people fear that without creation, the world becomes dull.

But the opposite is true.

A fixed creation story makes life smaller. It says each species was made as it is. It gives a quick answer and closes the book.

Evolution opens the book.

It shows that every living thing has a history. A bird is not just a bird. It carries the story of ancient reptiles, feathers, flight, survival, and extinction. A whale is not just a whale. It carries the story of land mammals returning to the sea. The human hand is not just a hand. It carries the story of vertebrate limbs, grasping, tool use, touch, and care.

Science does not remove wonder.

It gives wonder roots.

It allows us to see life not as a collection of finished products, but as a vast, unfolding history. That view is deeper than the old belief in separate creation. It is richer, more honest, and more moving.

The religious story says, "God made it."

Science asks, "How did it come to be?"

That second question takes us much farther.

The Calm Conclusion

Before Darwin, belief in a creator of life was understandable.

People saw design because they had no natural explanation for adaptation. They saw fixed species because they could not see deep time. They saw humans as separate because they did not yet understand our biological kinship with other life. They saw fossils and tried to fit them into sacred stories because they did not yet understand Earth's ancient history.

But science changed the question.

Fossils showed that life had a past.

Extinction challenged perfect creation.

Geology revealed deep time.

Classification hinted at nested order.

Comparative anatomy showed patterns of relationship.

Darwin gave the natural mechanism that made sense of it all.

Creation stories were early human attempts to explain life. They may still carry cultural, poetic, and emotional meaning, but they do not provide scientific evidence for God.

Once modern biology began to explain life naturally, the need for divine creation weakened.

That is the movement of this book:

Not from beauty to emptiness.

Not from meaning to despair.

But from myth to evidence.

From sacred story to natural history.

From God as explanation to science as understanding.

Chapter Summary

Before modern science, creation stories felt reasonable because people lacked natural explanations for life's diversity.

The idea of fixed species made separate creation appear obvious to earlier generations.

Religious views of life offered comfort, hierarchy, purpose, and human specialness.

Early classification named and organized living beings, but it did not explain their origin.

Fossils and extinction began to challenge the idea of a fixed, perfect creation.

Geology revealed deep time, making gradual biological change possible.

The similarities among living beings are better explained by common ancestry than by separate creation.

Creation stories may have cultural and emotional meaning, but they are not scientific evidence.

Before Darwin, God occupied a major gap in human understanding. Darwin would help close that gap.

Key Takeaway

Before Darwin, life seemed to need a creator because human beings had no strong natural explanation for adaptation, diversity, fossils, extinction, or human origins. Creation stories filled the gap. But as science advanced, that gap began to close. The living world no longer needed to be explained as separate divine creation. It could be understood as natural history.

Sources

Charles Darwin. On the Origin of Species.

Charles Darwin. The Descent of Man.

Charles Lyell. Principles of Geology.

William Paley. Natural Theology.

Georges Cuvier. Essay on the Theory of the Earth.

National Academies of Sciences, Engineering, and Medicine. Science, Evolution, and Creationism.

University of California Museum of Paleontology. Understanding Evolution: History of Evolutionary Thought.

Neil Shubin. Your Inner Fish.

Jerry A. Coyne. Why Evolution Is True.

Transition to the Next Chapter

The old creation view did not weaken because people suddenly stopped feeling religious.

It weakened because evidence accumulated.

But evidence still needed a unifying idea.

That idea came from a quiet English naturalist who spent years observing fossils, islands, barnacles, pigeons, plants, animals, and the small differences between living things.

His name was Charles Darwin.

And his discovery changed the meaning of life forever.

Chapter 4: Darwin's Dangerous Discovery

The Quiet Man Who Shook Creation

Charles Darwin did not look like a man who would shake the religious world.

He was not a fiery speaker. He was not an angry atheist trying to lead a public revolt against God. He was a careful English naturalist with patient eyes, a disciplined mind, and the courage to let evidence disturb inherited belief.

That is one reason his idea became so powerful.

Darwin did not weaken the old creation story by shouting at it.

He studied nature until the story no longer fit.

He looked at fossils, islands, birds, barnacles, plants, worms, pigeons, shells, and the small differences between living beings. He noticed patterns others had missed, or had not dared to follow fully. He saw that life was not fixed in the way many people had assumed. He saw variation everywhere. He saw that organisms seemed fitted to their surroundings, but not perfectly. He saw that living beings resembled one another in ways that suggested family history, not separate divine manufacture.

Slowly, a dangerous idea took shape.

Species were not separately created by God in their present form.

They changed.

They descended from earlier forms.

They were shaped by natural causes.

This was dangerous not because it made life less beautiful, but because it made divine creation unnecessary as a scientific explanation. Darwin did not merely add another fact to biology. He removed one of the oldest jobs religion had given to God.

Before Darwin, God was widely used to explain the diversity of life.

After Darwin, nature could explain it.

That is why his discovery still matters.

The Voyage That Changed His Mind

In 1831, Darwin joined the HMS Beagle as a young naturalist. The voyage lasted nearly five years and carried him across oceans, islands, forests, coastlines, mountains, and fossil-rich landscapes.

Darwin was young, but the living world became his classroom.

He saw tropical forests thick with life. He examined geological formations. He collected specimens. He observed animals and plants in different environments. He found fossils of extinct creatures in South America. Some of those fossils resembled living animals from the same region.

That mattered.

It suggested that life in one place might be connected to earlier life in that same place.

On islands, Darwin noticed another important pattern.

Island species often resembled species from nearby continents, but they were not identical. They looked related, yet changed. This was not what separate creation would naturally lead us to expect. If God created each species independently, why should island life so often resemble nearby mainland life? Why not create entirely different forms wherever desired?

The pattern made more sense if island species had descended from mainland ancestors and then changed over time.

The Galápagos Islands later became famous in this story. Darwin collected birds, tortoises, plants, and other specimens there. He did not understand everything immediately. Science rarely works like a sudden flash of light. Often, evidence becomes clear only after years of comparison, reflection, correction, and courage.

But the seeds were planted.

The old idea of fixed species had begun to crack.

Life Was Not Fixed

The religious view of life had long depended on fixed categories.

A dog was a dog.

A horse was a horse.

A bird was a bird.

A human was a human.

This matched ordinary experience. In one lifetime, we do not see a fish become a mammal or a reptile become a bird. We see parents and children. We see small differences. We see resemblance. But the larger categories seem stable.

Darwin began to understand that ordinary experience was too short to see the deep story.

Life changes across generations.

Large changes require deep time.

Human eyes are not naturally built to see millions of years.

We see the surface of life. Science allows us to see its history.

Darwin's insight was not that one creature suddenly transforms into another. That is a common misunderstanding. His insight was that small inherited differences, if filtered over long stretches of time, can produce major biological change.

That idea turned biology from a museum of fixed creations into a history of living transformation.

The living world was not a finished painting.

It was an unfolding story.

Variation: The Small Differences That Matter

Darwin saw that individuals within a species are not identical.

No two humans are exactly alike. No two dogs are exactly alike. No two pigeons, flowers, insects, or trees are exactly alike. Variation is everywhere.

Some variation may not matter much. Some may matter greatly.

One bird may have a slightly stronger beak. Another may see a little better. One plant may tolerate dryness more effectively. Another may resist disease more successfully. One animal may run slightly faster. Another may hide slightly better. One organism may leave many offspring. Another may leave none.

To many people, such small differences seem too weak to explain the diversity of life.

But Darwin saw their power.

Small differences are the raw material of selection.

If a difference helps an organism survive or reproduce, even slightly, that difference can become more common over generations. If the environment continues to favor it, the effect can grow. Across deep time, small advantages can accumulate into large changes.

This is difficult for some people to accept because the human mind is impressed by sudden events. We notice lightning, birth, death, injury, disaster, and dramatic change. Slow change is harder to feel.

But nature often works slowly.

A canyon is not carved in one storm.

A mountain is not worn down in one season.

A language does not change completely in one generation.

Life does not need to transform suddenly for evolution to be real.

Small changes, repeated long enough, can reshape the living world.

Artificial Selection: Darwin's Simple Clue

Darwin paid close attention to breeders.

Human beings had already changed animals and plants by selecting which individuals reproduced. Dogs, pigeons, cattle, sheep, horses, and crops showed that inherited traits could be shaped across generations.

This was a powerful clue.

A farmer saves seeds from the best plants. A breeder chooses animals with desired traits. Over time, the population changes. The result may look very different from the original form.

Consider dogs. A Great Dane, a dachshund, a bulldog, and a greyhound are all dogs, but their bodies differ greatly. Human selection produced enormous variety from inherited variation. No supernatural event was needed. Selection across generations was enough.

Darwin saw the larger implication.

If human selection can produce dramatic changes in a few generations, what could nature do over millions of years?

Nature does not select with intention. It does not choose as a farmer chooses. It does not prefer beauty, kindness, comfort, or fairness. But nature filters. Some organisms survive and reproduce more successfully than others. Some traits leave more descendants. Others fade.

Human selection is conscious.

Natural selection is blind.

But both can change populations.

This simple comparison helped Darwin show how life could be shaped without divine design.

The Struggle for Existence

Darwin also understood that nature produces more life than can survive.

A tree may release thousands of seeds. A fish may produce thousands of eggs. Insects may lay large numbers of eggs. Even among mammals, not every young animal survives to adulthood. If all offspring survived and reproduced, populations would grow without limit.

But they do not.

Most organisms die before reproducing.

They die from hunger, disease, injury, weather, predators, parasites, competition, and chance. Nature is not a peaceful garden arranged for everyone's comfort. It is a struggle for limited resources.

This struggle is not always violent in the visible sense. It can be quiet.

A seedling grows in shade and dies.

A bird fails to find enough food.

A young animal is slightly slower and is caught.

A plant cannot tolerate drought.

A bacterium without resistance dies when exposed to an antibiotic.

In such a world, even small advantages can matter.

A slightly better beak, root, immune response, camouflage, sense of smell, or tolerance to cold can affect survival and reproduction.

This is the setting in which natural selection works.

It is not sentimental.

It is not moral.

It is not guided by love.

It is simply the result of life producing more offspring than the environment can preserve.

Natural Selection: A Simple Idea With Deep Consequences

Darwin's central idea can be stated simply.

Living beings vary.

Some variation is inherited.

More organisms are born than can survive.

Some inherited differences help survival or reproduction.

Those differences become more common over generations.

That is natural selection.

The idea is simple, but its consequences are enormous.

Natural selection can explain why organisms fit their environments. It can explain why predators have traits useful for hunting, why prey have traits useful for escape, why flowers attract pollinators, why parasites adapt to hosts, why bacteria become resistant, why animals develop camouflage, and why organs can become complex over time.

This process has no mind.

It does not know the future.

It does not aim at perfection.

It does not protect the innocent.

It does not create justice.

It does not work for human happiness.

It does not need divine guidance.

It requires variation, inheritance, selection, and time.

That is why natural selection is so powerful. It explains the appearance of design without a designer.

For religious belief, this was a major loss.

God was no longer needed to explain why living beings fit their ways of life. Nature could do that.

Descent With Modification

Darwin used a phrase that still carries enormous power:

Descent with modification.

It means that living beings descend from earlier living beings, but they are not perfect copies. They inherit, but with change. Over time, populations can become different. Lineages can split. New species can arise. Old species can disappear.

This idea explains both similarity and difference.

Children resemble parents, but they are not identical. Species resemble related species, but they also differ. Larger groups of life resemble one another in deeper ways because they share more ancient ancestry.

This is why the bones in a human arm, a bat wing, a whale flipper, and a cat leg share an underlying pattern. They serve different functions, but they are related in structure. They are modified versions of inherited anatomy.

Separate creation can only say, "God chose to make them similar."

Evolution explains why they are similar.

They are family.

Descent with modification replaced the old picture of separate creation with a branching picture of life. Species were not isolated divine products. They were relatives.

The living world was not a collection of objects.

It was a family tree.

The Tree Replaced the Ladder

Before evolution, many people imagined life as a ladder.

God stood at the top. Below God came angels or spiritual beings. Then humans. Then animals. Then plants. Then the lower forms of existence. This picture placed humans near the top of reality and separated us sharply from the rest of nature.

Evolution replaced the ladder with a tree.

A tree has branches. Some branches grow. Some split. Some die. Some continue. No branch is the purpose of the whole tree. Each branch has its own history.

Human beings are not the top rung of a divine ladder.

We are one branch of life.

We are unusual, but not separate. We have language, culture, memory, science, art, and moral imagination. These are extraordinary features. But they do not require special creation. They are features of an evolved species with a remarkable brain and social life.

This was one reason Darwin's idea was so disturbing. It did not merely change the explanation of plants and animals. It changed the explanation of us.

If humans are part of nature, then the religious claim of special human creation is weakened.

If humans share ancestry with other life, then we are not a separate divine product.

If our bodies and minds have natural histories, then God is no longer needed to explain what we are.

Why Darwin's Idea Was Dangerous

Darwin's idea was dangerous because it touched the foundations of religious belief.

It challenged fixed species.

It challenged special creation.

It challenged the belief that human beings stood outside nature.

It challenged the idea that life's complexity required a divine designer.

It challenged the authority of sacred texts when they made claims about the natural world.

It challenged the comforting belief that the living world was arranged with humans at the center.

This is why the debate around evolution has never been only about fossils or finches. It is about identity. It is about authority. It is about whether human beings are willing to accept that nature can explain what religion once claimed for God.

Many religious people sensed the danger immediately.

If species were not separately created, then a major religious claim was wrong. If humans were animals with ancestors, then the doctrine of special human creation was shaken. If natural selection explained adaptation, then the design argument lost much of its force.

Darwin did not merely change biology.

He changed humanity's place in reality.

Darwin Was Not Alone

Darwin deserves his place in history, but good science does not depend on one heroic figure.

Alfred Russel Wallace independently developed a theory of natural selection. His work helped push Darwin to present his own ideas. Other thinkers before them had questioned fixed species or contributed to geology, classification, and natural history. Science is often a collective movement, even when one name becomes the symbol.

This matters because evolution is sometimes attacked as "Darwinism," as if it were a personal belief system built around one man.

That is wrong.

Evolution is not a religion of Darwin.

Darwin made a great discovery, but modern evolutionary biology rests on evidence from many fields: fossils, genetics, anatomy, embryology, biogeography, molecular biology, ecology, medicine, and direct observation.

Science does not ask us to worship Darwin.

It asks us to examine evidence.

Darwin's insight survived because evidence supported it, and later science strengthened it.

Darwin Did Not Know DNA

Darwin was brilliant, but he did not know everything.

He did not know DNA.

He did not know genes as we understand them today.

He did not know chromosomes.

He did not know modern mutation theory.

He did not know molecular biology.

He did not know the genetic code.

This matters because critics sometimes act as if any gap in Darwin's knowledge weakens evolution. That is not how science works.

Darwin began a revolution. Later science expanded it.

Genetics explained inheritance. DNA showed how biological information is copied and changed. Molecular biology allowed species to be compared at the genetic level. Fossils continued to fill in parts of life's history. Developmental biology revealed deep connections among living forms. Medicine showed evolution in action through antibiotic resistance, viral change, and disease.

Darwin did not have the whole picture.

But he saw the central pattern correctly.

Science does not depend on one man being perfect. It depends on evidence, correction, testing, and growth. Darwin's theory became stronger because later discoveries supported and refined it.

This is very different from religious dogma.

Religion often begins with a sacred conclusion and protects it. Science begins with evidence and revises itself.

That is why science progresses.

What Darwin Did Not Claim

It is also important to be fair about what Darwin did not claim.

Darwin did not claim that life has no beauty.

He did not claim that human beings have no value.

He did not claim that love is unreal.

He did not claim that morality does not matter.

He did not claim that every question in biology was solved.

He did not fully explain the first origin of life.

His central claim was more specific and more powerful:

Species change over time, and natural selection is a major mechanism by which they become adapted to their environments.

That claim was enough to transform biology.

Some people reject evolution because they think it steals meaning. But evolution does not tell us that life is meaningless. It tells us that life has a natural history. It tells us that living beings are connected. It tells us that the human body and mind belong to nature.

Meaning is not destroyed by truth.

Meaning becomes more honest when it no longer depends on illusion.

The Religious Retreat

When science explains something that religion once claimed, religion often retreats.

At first, God directly creates every species.

Then evidence shows that species change.

So God becomes the hidden guide behind evolution.

At first, God creates humans separately.

Then evidence shows human ancestry.

So God becomes the giver of a soul somewhere in the process.

At first, God designs each organ.

Then evolution explains adaptation.

So God becomes the mysterious source behind the laws of nature.

This movement is familiar. As evidence grows, God moves farther away from direct explanation.

This is the God of the gaps. God is placed where science has not yet given a complete answer. But as knowledge grows, the gap becomes smaller.

Darwin closed one of the largest gaps.

The diversity of life no longer required separate divine acts. Adaptation no longer required a designer. Human origins no longer required special creation.

God was pushed back.

Not by anger.

Not by rebellion.

By evidence.

The Emotional Shock of Common Ancestry

Common ancestry is difficult for many religious people because it changes the story of human identity.

If we are specially created by God, then we may feel set apart from the rest of life. We may feel chosen. We may feel spiritually superior. We may feel that the universe has us in mind.

If we are evolved animals, some people fear that we become ordinary, accidental, or meaningless.

But that fear misunderstands dignity.

We do not lose value because we share ancestry with other life. A child does not lose value because he resembles his parents. A person does not lose value because he belongs to a family. In the same way, humanity does not lose value because we belong to the family of life.

Common ancestry does not degrade us.

It connects us.

It tells us that our lungs, bones, eyes, blood, nerves, and cells carry a story older than civilization, older than religion, older than language, and older than humanity itself. We are not strangers in nature. We are nature becoming aware of itself.

That is not humiliation.

It is a deeper kind of wonder.

Darwin and the End of Special Creation

The idea of special creation was once simple and powerful.

God made species.

God made humans separately.

God designed organs.

God gave every creature its place.

Darwin did not disprove every possible idea of God. That would not be the most careful scientific claim. But Darwin did something devastating to the scientific need for God in biology.

He showed that life's diversity could be explained without divine creation.

That is the key point for this book.

Science-based atheism does not need to prove that no imaginable god could exist in any abstract philosophical sense. It needs to show that the living world does not provide scientific evidence for God and that natural explanations are sufficient where supernatural explanations were once used.

Darwin gave one of those great natural explanations.

After Darwin, the claim “God must have created species because there is no other explanation” was no longer credible.

There was another explanation.

A better one.

A testable one.

A natural one.

The Power of a Mechanism

The strength of Darwin’s theory was not merely that it said life changed. Others had suggested change before him.

The strength was that it offered a mechanism.

Natural selection explained how change could occur without intention.

This matters deeply.

A vague belief that life changed over time would not be enough. A scientific theory needs a process. Darwin gave biology that process.

Variation supplies differences.

Inheritance passes some differences forward.

Struggle filters them.

Time accumulates them.

Populations change.

Lineages split.

New species arise.

Old species disappear.

No divine hand is needed at any step.

Biology is more complex than this simple outline. Modern evolutionary theory also includes mutation, genetic drift, gene flow, recombination, sexual selection, developmental constraints, and environmental change. But Darwin's central insight remains one of the foundation stones.

Natural processes can create the appearance of design.

That insight strikes directly at the heart of the design argument.

Why This Discovery Still Matters

Some readers may wonder why Darwin still matters today. Biology has moved far beyond the nineteenth century.

Darwin still matters because he changed the question.

Before Darwin, people asked, “How did God design life?”

After Darwin, science asked, “How did life change through natural processes?”

That shift was enormous.

It allowed biology to become a true historical science. It connected fossils, anatomy, geography, classification, behavior, development, and later genetics. It gave scientists a framework that could be tested, corrected, and expanded.

Without evolution, biology is a pile of separate facts.

With evolution, biology becomes a coherent story.

This is why evolution is not a small chapter in biology. It is the foundation that makes modern biology understandable.

And if evolution explains the diversity and adaptation of life, then divine design is no longer needed as the scientific explanation.

The Calm Conclusion

Darwin’s discovery was dangerous because it replaced supernatural creation with natural history.

It showed that species are not fixed. They change.

It showed that living beings are connected by descent.

It showed that adaptation can arise without design.

It showed that human beings belong to nature.

It showed that biology does not need separate acts of divine creation.

Darwin did not remove wonder from life. He deepened it. He showed that every living being carries a history. A bird is not merely a bird. It is a descendant of earlier life. A flower is not merely a flower. It is part of a long evolutionary story. A human being is not a clay figure shaped by divine hands. A human being is a living branch of the ancient tree of life.

This truth may disturb religious certainty, but it does not diminish life.

It makes life more real.

The old story said, "God made each living thing."

Darwin's discovery said, "Life has changed through natural causes."

That was one of the great turning points in the human understanding of reality.

From that point forward, life no longer needed a creator as a scientific explanation.

Chapter Summary

Darwin did not set out to attack religion, but his discovery deeply weakened the idea of special creation.

His voyage on the HMS Beagle exposed him to fossils, islands, variation, and geographical patterns that challenged fixed species.

Darwin saw that small inherited differences could matter across generations.

Artificial selection showed that populations could change when traits were selected.

Natural selection explained how nature could shape adaptation without intention.

Descent with modification replaced separate creation with common ancestry.

Evolution changed the picture of life from a ladder to a branching tree.

Darwin was not alone, and evolution is not a belief system built around one man.

Darwin did not know DNA, but later genetics and molecular biology strengthened his central insight.

Darwin's discovery made divine special creation unnecessary as a scientific explanation.

Key Takeaway

Darwin's dangerous discovery was that life could explain itself through nature. Variation, inheritance, struggle, selection, and deep time could produce adaptation and diversity without divine design. Once this natural mechanism became clear, the old claim that God was needed to create each species lost its scientific force. Life remained beautiful, but it no longer required a creator.

Sources

Charles Darwin. On the Origin of Species.

Charles Darwin. The Descent of Man.

Charles Darwin. The Voyage of the Beagle.

Alfred Russel Wallace. Contributions to the Theory of Natural Selection.

University of California Museum of Paleontology. Understanding Evolution: History of Evolutionary Thought.

National Academies of Sciences, Engineering, and Medicine. Science, Evolution, and Creationism.

Ernst Mayr. What Evolution Is.

Jerry A. Coyne. Why Evolution Is True.

Richard Dawkins. The Blind Watchmaker.

Janet Browne. Charles Darwin: Voyaging.

Transition to the Next Chapter

Darwin's discovery gave biology a natural explanation, but the heart of that explanation deserves closer attention.

How can a blind process create useful organs, instincts, and adaptations?

How can nature shape life without planning it?

How can selection work without a selector?

The answer is natural selection: blind, but not weak.

Chapter 5: Natural Selection: Blind, But Not Weak

The Field After the Drought

A farmer walks through his field after a long, dry summer.

The soil is cracked. Many plants are brown and lifeless. Some have bent stems and dry leaves. But here and there, a few plants are still standing. They are not beautiful in any romantic sense. They are simply alive.

The farmer did not choose them. He did not inspect every root. He did not decide which plant deserved survival. He did not whisper instructions into the soil.

The drought did the filtering.

Plants with slightly deeper roots, thicker leaves, better water storage, or some other useful trait survived a little better. Plants without those advantages died before leaving seeds. If the survivors reproduce, the next generation may contain more of the traits that helped them endure dryness.

No mind selected them.

But selection still happened.

This is the heart of natural selection. It is not a person. It is not a divine judge. It is not nature making a conscious decision. It is the result of unequal survival and reproduction in a world where living beings vary and not all can survive.

This is why natural selection can be blind and still powerful.

It does not see.

It does not plan.

It does not love.

It does not hate.

It does not know the future.

It does not care who suffers.

Yet over many generations, it can shape life.

For centuries, people looked at living beings and imagined a divine selector behind them. The eye, the wing, the flower, the immune system, the spider web, and the human body all seemed too useful to exist without intention.

Darwin's great discovery was that usefulness does not require a divine mind.

It requires variation, inheritance, struggle, reproduction, and time.

The Simple Logic of Selection

Natural selection begins with a simple fact:

Living beings are not identical.

In any population, there is variation. Some animals run faster. Some plants tolerate dryness better. Some bacteria survive chemical exposure better. Some birds have stronger beaks. Some insects are harder to see. Some organisms resist disease more effectively.

Variation alone is not enough. Some of that variation must be inherited. If a useful trait cannot be passed to offspring in some form, natural selection cannot build on it.

The next fact is equally important:

More organisms are born than can survive.

A fish may release thousands of eggs. A plant may produce thousands of seeds. Insects may lay large numbers of eggs. Even among mammals, not every young animal reaches adulthood. Food is limited. Space is limited. Water is limited. Safety is limited. Disease, weather, injury, predators, parasites, and chance all affect survival.

This creates a natural filter.

If certain inherited traits help an organism survive and reproduce, those traits may become more common in future generations. If other traits reduce survival or reproduction, they may become less common.

That is natural selection.

The logic is simple, but its consequences are enormous:

Individuals vary.

Some differences are inherited.

More organisms are born than can survive.

Some inherited differences help survival or reproduction.

Those differences become more common over generations.

This process explains one of the greatest facts in biology: living beings can become fitted to their environments without being designed by God.

Variation: Nature's Raw Material

Without variation, natural selection would have nothing to work with.

If every plant in a field were genetically identical, and every plant responded to drought in exactly the same way, selection could not favor one inherited difference over another. If every bird had the same beak, every bacterium had the same resistance, and every insect had the same color, selection would have no raw material.

But life is full of difference.

Variation comes from mutation, recombination, sexual reproduction, gene flow, and other biological processes. Some variation is visible, such as height, color, shape, or behavior. Some is hidden in genes and proteins. Some matters greatly. Some matters very little. Some becomes important only when the environment changes.

This matters because religious critics often imagine evolution as if it had to create a full organ or a complete species in one sudden step.

That is not how evolution works.

Natural selection does not need a modern eye to appear all at once. It can begin with small differences in light sensitivity. It does not need a bird wing to appear fully formed. It can work with earlier structures that helped balance, warmth, display, gliding, or movement. It does not need a perfectly adapted organism. It works with small inherited differences already present in a population.

Evolution does not begin with perfection.

It begins with variation.

Inheritance: How Differences Continue

A useful difference matters to evolution only if it can influence future generations.

Suppose one animal becomes stronger because it exercised more during life. That strength may help that individual, but if the change is not inherited, natural selection cannot directly build on it in the next generation.

Inherited differences are different.

If a bird is born with a beak shape that helps it crack available seeds, and if that trait has a genetic basis, its offspring may inherit a similar advantage. If a bacterium has a genetic change that helps it survive an antibiotic, its descendants may carry that resistance. If a plant has inherited traits that help it tolerate salty soil, its seeds may pass some of that advantage forward.

Inheritance gives natural selection memory.

Not conscious memory.

Biological memory.

It allows successful traits to persist beyond one life. Across generations, this can change the character of a population.

This is one reason evolution is not random chaos. Each generation begins with what was inherited from earlier generations. Life carries its past into its future.

A divine designer is not needed to explain this.

Inheritance and selection are enough.

The Struggle for Existence

The word struggle can sound harsh, but it is accurate.

Life produces more offspring than the environment can support. Not every seed becomes a tree. Not every egg becomes an adult. Not every young animal survives. Not every organism finds food, avoids disease, escapes predators, and reproduces.

This struggle is not always dramatic. It is not always claws, teeth, and blood.

It may be a seedling dying in shade.

It may be a bird failing to find food.

It may be a young animal that is slightly slower than the others.

It may be a plant that cannot survive frost.

It may be a bacterium killed by a drug.

It may be an insect eaten because its color made it more visible.

Life is constantly filtered by conditions.

This is not moral judgment. It is not reward and punishment. It is not divine choice. It is simply the fact that some organisms leave more descendants than others.

The living world is not arranged around fairness.

It is shaped by survival and reproduction.

That hard truth is central to evolution.

It is also one reason nature does not look like the work of a loving designer.

Selection Without a Selector

The word selection can mislead us.

When we select something, we choose. We select a shirt, a meal, a book, a medicine, or a route. Selection usually implies a mind.

Natural selection is different.

There is no conscious selector standing outside life. No divine hand is choosing which beak, tooth, claw, root, wing, or immune response should continue. No invisible judge is deciding which creature deserves survival.

The environment filters.

Predators filter.

Climate filters.

Food supply filters.

Disease filters.

Competition filters.

Mating success filters.

Chance events filter.

Organisms that survive and reproduce leave more descendants. Their traits become more common, not because nature loved them, but because they persisted.

This is selection without intention.

It may feel strange at first because human beings prefer purposeful explanations. We are used to asking, "Who did it?" Who made the eye? Who designed the wing? Who planned the flower? Who shaped the body?

Natural selection changes the question.

It asks something better:

What traits helped survival or reproduction?

How were those traits inherited?

What conditions favored them?

How did the population change over generations?

These are scientific questions.

They can be studied.

They do not require God.

Evolution Is Not Pure Chance

One of the most common religious objections says, "Evolution cannot be true because life could not arise by random chance."

This objection attacks a false version of evolution.

Evolution is not pure random chance.

Mutation and genetic variation may arise without regard to what an organism needs. A bacterium does not decide to mutate because an antibiotic is coming. A plant does not create a useful genetic change because a drought is expected. Variation is not guided by future purpose.

But natural selection is not random.

Selection filters variation. If a trait helps survival or reproduction in a particular environment, organisms with that trait may leave more descendants. If a trait is harmful, organisms carrying it may leave fewer descendants.

So the process has two sides:

Variation supplies possibilities.

Selection filters them.

This is why evolution can produce organized adaptation without divine planning. It is not like a storm assembling a complete airplane from junk. That old comparison is deeply misleading. Evolution does not assemble a full organism in one random event. It works through small inherited differences across many generations, with selection preserving what works well enough.

A better comparison is a sieve.

Many variations pass through life. Some are lost. Some survive. Some become common. The filter is not conscious, but it is not meaningless.

Natural selection is blind.

But it is not helpless.

Fitness Does Not Mean Strength

Another common misunderstanding concerns the phrase "survival of the fittest."

Many people hear this and imagine the strongest animal winning every contest. They picture the largest male, the fiercest predator, the fastest runner, or the toughest fighter.

But in evolution, fitness does not simply mean strength.

Fitness means reproductive success in a particular environment.

A small animal may be more fit than a large one if it survives better and leaves more offspring. A quiet strategy may be more successful than an aggressive one. A plant with smaller leaves may be more fit in a dry climate. A bacterium with antibiotic resistance may be more fit in the presence of that antibiotic, but not necessarily in every environment.

Fitness is not moral superiority.

Fitness is not physical perfection.

Fitness is not intelligence.

Fitness is not virtue.

It is reproductive success under specific conditions.

This point matters because nature does not reward goodness. It does not reward kindness. It does not reward justice. It preserves whatever traits help genes continue in a given environment.

A parasite may be fit. A virus may spread successfully. A predator may thrive. A trait may persist even if it causes suffering.

This is not a moral system.

It is a natural process.

And that is exactly why it does not look like the work of a loving divine ruler.

Good Enough, Not Perfect

Natural selection does not produce perfection.

It produces what is good enough to survive and reproduce.

This is one of the most important differences between evolution and divine design. A perfect designer, especially one described as all-knowing and all-powerful, could build ideal structures. Evolution cannot. Evolution works with what already exists.

It modifies older parts.

It adjusts inherited systems.

It accepts compromises.

It preserves traits that work well enough.

It does not start over with a clean plan.

This is why living bodies are full of imperfections.

The human spine supports upright walking, but it also causes pain. The eye sees well, but it has a blind spot. The throat allows speech, but it also creates a risk of choking. The immune system protects us, but it can attack our own tissues. Cell division allows growth and repair, but it can become cancer.

These are not signs of perfect design.

They are signs of evolutionary compromise.

Natural selection does not ask, “What is the best possible body?” It filters among available variations. If a trait helps enough, it may persist even if it carries costs. If a structure works adequately, it may remain even if a thoughtful engineer could imagine something better.

This is why life often looks impressive from a distance and messy up close.

It is not designed perfection.

It is selected survival.

No Future Plan

Natural selection cannot see the future.

This is another major difference between evolution and design.

A human engineer can plan for later use. An architect can design a foundation for a building that does not yet exist. A physician can prescribe treatment to prevent a future complication. A parent can save money for a child's education.

Minds can imagine future states.

Natural selection cannot.

It can only filter present variation under present conditions.

A trait is not preserved because it might help a species a million years later. It is preserved if it helps survival or reproduction now, or at least does not prevent it. This is why evolution can produce dead ends. A trait that works well in one environment can become harmful when conditions change. A species highly adapted to one way of life may become vulnerable when that way of life disappears.

This explains much of life's fragility.

It also explains extinction.

A perfect designer could plan for every future.

Evolution cannot.

Life survives as long as its inherited traits remain good enough under changing conditions.

Natural selection is powerful, but it is blind to tomorrow.

Trade-Offs Are Everywhere

Biology is filled with trade-offs.

A strong immune response may protect against infection, but it can also cause damaging inflammation or autoimmune disease. Blood clotting prevents fatal bleeding, but it can also cause heart attack, stroke, or pulmonary embolism. Anxiety may help detect danger, but it can become disabling. Cell division allows growth and repair, but it increases the risk of cancer. Upright walking freed human hands, but it contributed to back pain, knee problems, and difficult childbirth.

A perfect designer could avoid cruel trade-offs.

Evolution often cannot.

Natural selection preserves traits that improve survival or reproduction on balance, not traits that are perfect in every way. A feature may help early in life but cause problems later. A trait may help in one environment but harm in another. A system may solve one problem while creating another.

This is why the body is not a flawless machine.

It is a collection of workable compromises.

Trade-offs are not embarrassing details for evolution.

They are expected.

But they are deeply uncomfortable for the idea of perfect design.

Why Harmful Traits Can Persist

Some people ask, "If natural selection removes harmful traits, why do genetic diseases and biological weaknesses still exist?"

The answer is that selection is powerful, but not perfect.

Some harmful mutations appear again and again. Some cause problems only later in life, after reproduction has already occurred. Some are recessive and can be carried silently by healthy people. Some persist because they are linked to other traits. Some may carry an advantage in one setting and a disadvantage in another. Some remain because chance also plays a role in evolution.

This explains why bodies are not purified into perfection.

Natural selection is not a divine quality-control system.

It does not remove every error.

It does not protect every child.

It does not prevent every disease.

It does not create the best possible genome.

It filters, but it filters imperfectly.

This is why genetic disease is not surprising in evolution. It is expected in a world of mutation, inheritance, reproduction, selection, and biological limits.

But it is troubling for perfect design.

A perfect designer could avoid harmful mutations.

Natural selection cannot.

Complexity Can Grow Step by Step

Another religious objection says that complex organs could not evolve because they are useless until complete.

This objection misunderstands evolution.

Complex structures can evolve through intermediate stages if each stage provides some advantage.

An eye does not need to begin as a modern human eye. A simple light-sensitive patch can help an organism detect light and dark. A shallow cup can help detect direction. A deeper cup can improve directional sensitivity. A transparent covering can protect delicate cells. A lens-like structure can improve focus. Each stage can be useful.

A wing does not need to begin as a fully formed flight organ. Earlier structures may help with balance, insulation, display, gliding, or maneuvering before full flight evolves.

A complex immune system does not need to appear in one step. Defensive systems can become more layered and specialized over time.

Evolution does not require a finished organ to appear in one impossible jump.

It requires useful steps.

This is where the design argument often fails. It looks at the final structure and says, "This could not have appeared all at once."

Evolution agrees.

It did not appear all at once.

It developed through earlier forms, each shaped by natural processes.

Antibiotic Resistance: Selection in the Clinic

Natural selection is not only something that happened in ancient forests or distant oceans.

It happens in hospitals and clinics.

Antibiotic resistance is one of the clearest examples.

In a population of bacteria, some may carry genetic traits that allow them to survive exposure to an antibiotic. When the antibiotic is used, many susceptible bacteria die or stop multiplying. Resistant bacteria survive. They reproduce. Over time, the resistant bacteria become more common.

The bacteria did not plan this.

They did not know the antibiotic was coming.

The antibiotic changed the environment. Resistant variants had an advantage in that environment. They survived and multiplied.

That is natural selection.

This is not a philosophical debate. It is a medical reality.

Doctors must think about resistance when choosing antibiotics. Hospitals track resistant organisms. Public health experts warn against unnecessary antibiotic use because overuse increases selection pressure. Resistant germs are not evidence of divine design. They are evidence of evolution in action.

The religious claim that life is designed by God adds nothing useful here.

It does not help choose treatment.

It does not help prevent resistance.

It does not explain bacterial genetics.

Evolution does.

Sexual Selection: Beauty Without Divine Art

Natural selection is not the only evolutionary process that shapes living beings. Sexual selection also matters.

Some traits spread because they improve mating success rather than direct survival. Bird songs, bright colors, courtship dances, antlers, displays, ornaments, and mating behaviors may be shaped this way.

These features can look like divine decoration.

But beauty in nature does not require a divine artist.

It can arise from mate choice, competition, signaling, and reproductive success.

A peacock's tail may be beautiful, but it can also be costly. It requires energy and may attract predators. Antlers can impress or help in competition, but they can also injure. Bright colors may help attract mates, but they can reduce camouflage.

Beauty and burden often come together.

This is exactly what evolution would predict.

A trait may spread if it improves reproduction, even if it carries costs.

Perfect design is not needed.

Evolution Is Not a Ladder of Progress

Another misunderstanding says that evolution is a climb from lower life to higher life, ending with human beings.

This is false.

Evolution is not a ladder.

It is a branching process.

Bacteria are not failed humans. Fish are not trying to become mammals. Monkeys are not waiting to become people. Every living species has its own history of survival. Some lineages become more complex in certain ways. Some remain simple because simplicity works. Some disappear.

There is no built-in goal toward humanity.

This matters because religion often imagines humans as the purpose of creation. Evolution does not support that view. Human beings are one branch of life, not the final aim of the whole process.

If ancient events had unfolded differently, humans might never have appeared. Different extinctions, climate shifts, mutations, migrations, and accidents could have changed everything.

This does not make human life meaningless.

It makes human life contingent.

We are not the guaranteed product of a divine plan.

We are one remarkable outcome of natural history.

The Power of Repetition Across Time

Natural selection seems too simple to some people.

They ask how such a basic process could shape the living world.

The answer is repetition across time.

One generation may change little. Ten generations may change a little more. A thousand generations can show significant change. A million generations can transform a lineage.

This is easier to understand in organisms that reproduce quickly. Bacteria can pass through many generations in a short time. Insects, plants, and small animals can also show measurable change under strong selection pressures.

Larger animals usually change more slowly, but deep time allows even slow processes to matter.

The human mind struggles with deep time because our lives are short. We think in years, decades, and centuries.

Evolution often works in thousands, millions, and billions of years.

A small difference repeated across immense time is not small.

It is world-changing.

Natural selection does not need sudden creation.

It needs repeated filtering across generations.

Why Natural Selection Weakens Divine Design

Natural selection does not disprove every possible idea of God. A person can always move God into an invisible place beyond testing.

But natural selection strongly weakens the claim that life provides scientific evidence of divine design.

Why?

Because it gives a natural mechanism for adaptation.

Before Darwin, adaptation was the strongest argument for God in biology. The eye seemed designed for seeing. The wing seemed designed for flying. Teeth seemed designed for chewing. Flowers seemed designed for pollinators.

After Darwin, these features could be explained without divine intention.

This does not mean belief disappears immediately. Belief can survive by becoming less testable. But as a scientific explanation, God becomes unnecessary.

A hidden designer who cannot be detected, measured, tested, or distinguished from natural selection adds nothing to biology.

Natural selection explains the pattern.

It explains why organisms are fitted but not perfect.

It explains why adaptation is local and temporary.

It explains why bodies carry compromises.

It explains why harmful traits can persist.

It explains why species change.

It explains why extinction happens.

It explains why life is beautiful, brutal, and indifferent.

That explanatory power is why natural selection is so important for science-based atheism.

It removes the need for a divine designer in biology.

A Hard Truth With a Human Meaning

Natural selection can be emotionally difficult to accept because it does not flatter us.

It does not say life was made for our comfort.

It does not say humans were planned from the beginning.

It does not say nature is guided by love.

It does not say every suffering has a purpose.

It tells a harder truth.

Life has been shaped by survival and reproduction in a changing world. Many organisms died. Some reproduced. Their descendants carried forward traits that worked well enough. Over time, life diversified, adapted, struggled, and disappeared in countless forms.

This is not a comforting story in the religious sense.

But it is honest.

And honesty has its own dignity.

A scientific view of life does not require us to pretend that nature is kind. It allows us to see nature clearly. Once we see clearly, we can build human values not from illusion, but from compassion.

Because nature is indifferent, human kindness matters more.

Because evolution is blind, human moral vision matters more.

Because there is no divine plan protecting life, our responsibility to one another becomes greater, not smaller.

The Calm Conclusion

Natural selection explains how life can appear designed without being designed.

It is not random chaos. It is not magical progress. It is not a divine tool guided from heaven. It is a natural process that acts whenever inherited variation affects survival and reproduction across generations.

It is blind, but it is not weak.

It has shaped wings, eyes, roots, flowers, claws, immune systems, instincts, and bodies. It has also left suffering, imperfection, disease, waste, and extinction.

That mixture is exactly what we should expect from evolution.

It is not what we should expect from perfect divine design.

The living world does not show the signature of an all-loving engineer.

It shows the marks of a powerful natural filter.

No divine selector is required.

Chapter Summary

Natural selection is the result of unequal survival and reproduction among organisms with inherited variation.

It has no mind, intention, mercy, or future plan.

Evolution is not pure random chance. Variation may arise without regard to need, but selection filters variation in non-random ways.

Fitness means reproductive success in a specific environment, not strength, moral worth, or perfection.

Natural selection produces adaptation, not perfect design.

Bodies are often good enough to survive and reproduce, but they remain full of flaws and trade-offs.

Harmful traits can persist because selection is powerful but imperfect.

Complex structures can evolve through useful intermediate stages.

Antibiotic resistance shows natural selection in a medical setting.

Natural selection weakens the design argument because it explains adaptation without God.

Key Takeaway

Natural selection is blind, but it is powerful. It can shape life without planning it, preserve useful traits without loving them, and create the appearance of design without a designer. Once natural selection is understood, the old claim that life must have been designed by God loses its scientific force. Life does not require a divine selector. Nature itself does the filtering.

Sources

Charles Darwin. On the Origin of Species.

Theodosius Dobzhansky. Nothing in Biology Makes Sense Except in the Light of Evolution.

University of California Museum of Paleontology. Understanding Evolution: Natural Selection.

University of California Museum of Paleontology. Understanding Evolution: Adaptation.

National Academies of Sciences, Engineering, and Medicine. Science, Evolution, and Creationism.

Centers for Disease Control and Prevention. About Antimicrobial Resistance.

Ernst Mayr. What Evolution Is.

Jerry A. Coyne. Why Evolution Is True.

Richard Dawkins. The Blind Watchmaker.

Sean B. Carroll. Endless Forms Most Beautiful.

Transition to the Next Chapter

Natural selection explains how evolution works. But some readers may still think evolution is only an idea about the distant past.

The next chapter brings evolution into the present.

We will see that evolution is not merely ancient history written in fossils. It is happening now, in bacteria, viruses, insects, plants, animals, and even in the medical problems we face today.

Evolution is not dead history.

It is life changing before our eyes.

Chapter 6: Evolution in Real Time

The Bacteria That Survived

A patient comes to the hospital with a serious infection.

At first, the plan seems clear. A culture is sent. An antibiotic is started. The fever should come down. The infection should begin to improve. The family waits for the medicine to work.

Sometimes it does.

But sometimes the story becomes more troubling. The fever continues. The patient remains weak. The laboratory report returns with words no physician likes to see:

Resistant organism.

The antibiotic that should have worked does not work. A stronger medicine may be needed. Sometimes even that is difficult. The bacteria have changed, and treatment must change with them.

This is not a religious mystery.

It is not punishment from heaven.

It is not a supernatural event.

It is evolution.

In the clinic, evolution is not an abstract idea from a biology textbook. It can become a matter of life and death. Bacteria reproduce quickly. Mutations occur. Some bacteria carry traits that allow them to survive exposure to antibiotics. When an antibiotic kills the susceptible bacteria, resistant ones may remain. They multiply. The infection becomes harder to treat.

This is natural selection in real time.

No divine hand is needed. No invisible designer is guiding the bacteria. No supernatural mind is deciding which organisms should live. The process is natural, observable, and medically important.

Many people think evolution is only about ancient fossils, dinosaurs, or distant events no one can see.

That is not true.

Evolution is happening now.

It is happening in hospitals.

It is happening in farms.

It is happening in insects.

It is happening in viruses.

It is happening in plants.

It is happening wherever living populations vary, reproduce, and face pressure from their environment.

The living world is not fixed.

It is changing before our eyes.

Evolution Is Not Only Ancient History

Many objections to evolution depend on distance.

The critic says, “No one was there to see it.”

The critic says, “Evolution happened too long ago to prove.”

The critic says, “It is only a theory about the past.”

But evolution is not only a story about the distant past. It is also a process we can observe in the present.

This matters.

If the same natural mechanism we see today can change living populations, then we have no reason to treat ancient evolution as fantasy. The present helps us understand the past. Much of science works this way.

We do not need to watch a mountain form from beginning to end to understand geology. We study erosion, volcanic activity, sedimentation, pressure, uplift, and rock formation today.

We do not need to watch a star live its entire life to understand astronomy. We study light, gravity, fusion, stellar patterns, and different stages of stars across the universe.

We do not need to witness every ancient event directly to know that natural processes leave evidence.

Evolution works in the same way.

We can observe variation.

We can observe inheritance.

We can observe selection.

We can observe changes in populations.

We can observe resistance.

We can observe adaptation.

The fact that evolution also happened across deep time does not make it unscientific. It means that present processes, combined with fossils, DNA, anatomy, geography, and development, help us reconstruct life's long history.

Evolution is not belief without evidence.

It is a scientific explanation supported by many lines of evidence, including changes we can watch today.

What It Means to Observe Evolution

To observe evolution, we do not need to see a fish turn into a mammal before our eyes.

That is not how evolution works.

Evolution means change in inherited characteristics of populations over generations.

The word population is important. Evolution does not mean that one individual suddenly transforms into a different kind of creature. A bacterium does not become a bird. A monkey does not give birth to a human. These are common misunderstandings, often repeated by people who do not describe evolution accurately.

Individuals are born, live, and die.

Populations evolve.

When inherited traits become more or less common over generations, evolution has occurred.

This can happen quickly in organisms that reproduce rapidly. Bacteria, viruses, insects, and some plants can show noticeable evolutionary change in short periods. Larger animals usually change more slowly, but if selection pressure is strong enough, measurable changes can still occur.

The speed of evolution depends on several things:

How quickly the organism reproduces.

How much variation exists.

How strongly the environment favors one trait over another.

How isolated the population is.

How many generations pass.

Small changes observed today are not separate from large evolutionary changes across deep time. They are the same kind of process working over different time scales.

A single step is not a journey.

But every journey is made of steps.

Antibiotic Resistance: Evolution in Medicine

Antibiotic resistance is one of the clearest examples of evolution in real time.

Imagine a large population of bacteria. Most are sensitive to a certain antibiotic. A few carry genetic traits that allow them to survive. When the antibiotic is given, many sensitive bacteria die or stop multiplying. Resistant bacteria survive. They reproduce. Soon the bacterial population contains a much higher proportion of resistant organisms.

This is not speculation.

It is a daily reality in modern medicine.

Doctors must consider resistance patterns when choosing antibiotics. Hospitals track resistant organisms. Public health experts warn against unnecessary antibiotic use because overuse increases selection pressure. The more often bacteria are exposed to antibiotics, the more opportunity resistant strains have to survive and spread.

The antibiotic does not teach bacteria how to resist.

The bacteria do not think.

They do not plan.

They do not pray for survival.

Some already have, or acquire, traits that help them survive. The antibiotic changes the environment. The resistant bacteria survive better in that environment. Their descendants become more common.

That is natural selection.

A divine design explanation adds nothing useful here. It does not help us predict resistance. It does not help us choose treatment. It does not help us understand bacterial genetics. It does not help infection control.

Evolution does.

This is why evolution is not optional in modern biology. It is not merely a philosophical position. It is a working tool in medicine.

The Physician's View

A physician cannot honestly look at infectious disease without seeing the importance of evolution.

Every antibiotic prescription exists in an evolutionary world. Every unnecessary antibiotic, every incomplete treatment, every agricultural overuse of antimicrobial drugs, every hospital outbreak, and every resistant strain reminds us that bacteria are not fixed targets.

They change.

This does not mean a physician thinks about Darwin every time a prescription is written. But the principle is always there. Microbial populations respond to pressure. Resistant organisms survive. Populations shift. Medicine must adapt because life adapts.

This is why supernatural explanation is not merely unnecessary. It is unhelpful.

If we say, "God designed bacteria," what have we explained?

If we say, "God allowed resistance," what have we predicted?

If we say, "It is part of God's plan," what treatment decision follows?

Nothing useful follows.

But if we understand mutation, selection, resistance genes, transmission, and environmental pressure, we can act. We can test. We can isolate. We can prescribe carefully. We can improve infection control. We can reduce unnecessary antibiotic use. We can develop new drugs and new strategies.

Science saves lives because it explains mechanisms.

Faith may comfort a patient emotionally.

But it does not explain antibiotic resistance.

Evolution does.

Pesticide Resistance: The Same Logic in the Field

The same process occurs outside medicine.

Farmers may use pesticides to control insects or other pests. At first, the pesticide may work well. Many insects die. But if a few insects carry traits that allow them to survive, they reproduce. Over generations, the population may become more resistant.

The pesticide becomes less effective.

The insects do not decide to become resistant. They do not understand chemistry. They do not develop a strategy. There is variation in the population. The pesticide creates strong selection pressure. Survivors reproduce. Resistance becomes more common.

Different setting.

Same evolutionary principle.

This shows that evolution is not limited to one type of organism. Whenever inherited variation affects survival and reproduction, populations can change.

A religious explanation cannot improve this understanding. Saying "God designed insects" does not explain why pesticide resistance increases. Saying "God gave them survival ability" does not guide agricultural science.

Evolution does.

It helps farmers and scientists understand why rotating control methods, reducing unnecessary chemical pressure, using integrated pest management, and protecting ecosystems can matter.

The world behaves as if natural selection is real.

Because it is.

Viruses and Rapid Change

Viruses also show how quickly evolution can matter.

Viruses replicate. During replication, genetic changes can occur. Some changes do little. Some harm the virus. Some may help it spread, enter cells, avoid parts of immune recognition, or persist in a population under certain conditions.

Because many viruses reproduce rapidly, evolution can occur quickly.

This is why scientists monitor viral variants. They compare genetic sequences. They track changes over time. They study whether a variant spreads more easily, responds differently to immunity, or changes the pattern of disease.

Influenza is a familiar example. Flu viruses change over time, which is one reason vaccines may need updating. HIV also evolves rapidly, which contributes to treatment challenges and the need for combination therapy. Many other viruses show the same basic principles: replication, variation, selection, and spread.

This does not mean viruses are thinking.

They are not planning.

They are not trying to defeat us in a conscious way.

They replicate. Variation appears. Host immunity, medication, transmission opportunities, and chance help determine which variants continue.

Once again, no divine guidance is needed.

The more carefully we study living systems, the more we find processes, not miracles.

Laboratory Evolution: Watching Change Under Controlled Conditions

Evolution can also be studied in the laboratory.

Scientists can grow bacteria or other organisms over many generations and observe how populations change under specific conditions. In these experiments, researchers can control the environment, preserve samples, compare earlier and later generations, sequence genomes, and measure changes in fitness.

This is powerful because it allows evolution to be tested directly.

If a bacterial population is placed in a particular environment for many generations, will it adapt? Will genetic changes appear? Will some lineages survive better than others? Can earlier and later generations be compared?

The answer is yes.

Laboratory evolution shows that populations can change in measurable ways. It shows that adaptation can occur without divine intervention. It shows that natural processes can produce new traits, new abilities, and new patterns of survival.

These experiments do not claim to reproduce the entire history of life in a laboratory. That is not the point. The point is that the mechanism works.

Populations vary.

Selection acts.

Genetic change accumulates.

Descendants differ from ancestors.

This is exactly the kind of evidence science values.

It is observable.

It is testable.

It is measurable.

It is open to correction.

The claim “God designed life” has none of these scientific advantages.

The Beak That Changed With the Weather

One of the best-known examples of evolution in wild animals comes from island finches.

In some finch populations, beak size and shape affect what food birds can eat. During dry periods, when small soft seeds become scarce and harder seeds remain, birds with stronger or deeper beaks may survive better. When conditions change, different beak traits may be favored.

The population can shift.

This does not mean a bird suddenly becomes a different kind of animal. It means inherited traits become more or less common because they affect survival and reproduction.

That is evolution.

This example is important because it helps remove the cartoon version of evolution. Critics sometimes demand absurd evidence, as if evolution requires one modern species to give birth suddenly to another modern species. But evolution does not predict that.

Evolution predicts population change across generations.

That is exactly what such examples show.

A small change in average beak size may seem modest. But repeated changes across long periods can contribute to larger differences among lineages.

Small change is not a weakness of evolution.

Small change is how evolution begins.

The Moth Against the Tree

Industrial melanism in moths has long been used as a simple example of natural selection.

In some environments, moth color affects visibility to predators. If tree bark becomes darker because of pollution, darker moths may be less visible and survive better. If the environment becomes cleaner and lighter surfaces return, lighter moths may regain an advantage.

The real history has details that scientists continue to study carefully, but the principle is clear:

Environmental conditions can change which traits are favored.

A moth does not choose its color to match the tree. A population contains variation. Predators notice some moths more easily than others. The more visible moths are eaten more often. The less visible moths survive more often and reproduce. Over generations, the population changes.

Again, the result can look designed.

The process is not design.

It is selection.

The moth against the tree teaches a simple lesson: survival may depend on the relationship between inherited traits and environment. Change the environment, and the value of a trait can change.

This is exactly what evolution predicts.

Color, Sand, and Survival

Another clear example comes from small animals whose coat color matches their surroundings.

In some regions, light-colored mice living on light sand are harder for predators to see. In nearby dark lava flows, darker mice may have an advantage. Over generations, natural selection can favor different coat colors in different environments.

This example is easy to understand.

Predators hunt by sight.

Some animals are easier to see.

The easier ones are more likely to be eaten.

The harder ones are more likely to survive.

Survivors leave more offspring.

The population changes.

No designer is needed to paint the mice.

The environment and predators filter the variation.

This example also shows why adaptation is local. A light coat may help on pale sand and harm on dark lava. A dark coat may help on lava and harm on pale sand.

A perfect designer might create universal perfection.

Evolution creates local fit.

That is exactly what we observe.

Evolution in Plants

Plants also evolve in observable ways.

Some plants become resistant to herbicides. Some adapt to heavy metals in polluted soils. Some change flowering time in response to climate, season, or pollinator pressures. Some develop traits that help them survive drought, salt, grazing, or disease.

Because plants cannot run away, their survival depends heavily on inherited traits that help them tolerate local conditions.

If a plant survives herbicide exposure and produces seeds, its descendants may carry resistance. If a plant cannot survive, it leaves fewer descendants or none. Over time, resistant plants may become more common.

The same evolutionary pattern appears again:

Variation.

Inheritance.

Selection.

Reproduction.

Change.

The more examples we consider, the clearer the pattern becomes. Evolution is not a narrow idea invented to challenge religion. It is a broad, practical explanation for living change.

It applies to bacteria, viruses, insects, birds, mammals, plants, and many other forms of life.

A divine explanation is not needed.

Artificial Selection Shows the Power of Selection

Artificial selection is guided by human intention, but it helps us understand the power of selection across generations.

Humans have changed dogs, pigeons, horses, cattle, crops, and many other living things by choosing which individuals reproduce. Corn looks very different from its wild ancestors. Dog breeds show enormous variation in size, shape, coat, behavior, and physical ability.

This does not prove natural selection by itself, because the selector is human.

But it demonstrates a key point:

Inherited traits in populations can change dramatically when selection is applied.

In artificial selection, the selector is a person.

In natural selection, there is no mind. The filter is the environment, survival, reproduction, competition, disease, climate, predators, and many other pressures.

Religious critics sometimes say selection cannot create meaningful change. Artificial selection shows that selection can reshape living populations. Nature has far more time than any breeder.

If human beings can produce dramatic changes in a few centuries, what can natural selection do over millions of years?

The answer is written in the living world.

"Only Microevolution" Is Not Enough

Some religious critics accept small evolutionary changes but reject larger ones.

They say, "Microevolution happens, but macroevolution does not."

This sounds like a compromise, but it often hides a misunderstanding.

Microevolution refers to smaller changes within populations. Macroevolution refers to larger patterns of evolutionary change, including speciation and major differences over long periods.

But macroevolution is not a different magic process. It is evolutionary change seen across larger scales, longer time, branching lineages, extinction, and reproductive separation.

A person who accepts that populations can change, that traits can be inherited, that selection can act, and that lineages can become isolated has already accepted the basic machinery of evolution.

If those processes continue long enough, larger differences can arise.

Of course, not every small change becomes a major transformation. Many changes remain limited. Many lineages disappear. Evolution does not guarantee progress.

But the divide between small change and large change is not a wall placed in nature by God.

It is a difference in scale.

A stream can cut a small groove in mud after one rain. Given enough time, water can carve a canyon. The small process and the large result are connected.

The same is true in evolution.

Speciation: When Populations Split

Evolution in real time also helps us understand speciation, the formation of new species.

Species do not usually appear suddenly. They often begin when populations become separated. The separation may be geographic, ecological, behavioral, or genetic. Over time, separated populations may accumulate differences. If those differences become large enough, the populations may no longer interbreed successfully.

At that point, we may recognize them as separate species.

Nature does not always draw sharp lines. Some species are clearly distinct. Others are connected by hybrids, ring populations, or gradual variation. This complexity is not a problem for evolution. It is exactly what we should expect if species are historical populations rather than fixed divine categories.

Creation thinking often imagines species as clean boxes.

Biology shows something more fluid.

Life branches.

Lineages split.

Boundaries can be messy because evolution is a process, not a filing cabinet.

This is another reason the old idea of fixed creation fails. Nature does not behave as if every species were separately placed in a permanent category. It behaves as if living populations have histories.

Why Observed Evolution Matters for This Book

Observed evolution matters because it removes one more hiding place for God.

If evolution were only a speculative story about the distant past, religious critics might dismiss it more easily. But when evolution can be observed in antibiotic resistance, viral change, pesticide resistance, laboratory experiments, plant adaptation, animal color changes, and shifts in wild populations, the situation changes.

Evolution becomes visible.

Once evolution is visible, the claim that life is fixed by divine creation becomes much weaker.

The living world does not behave like a completed set of products. It behaves like a changing system.

This does not mean every question is answered. Science never claims that. But the evidence clearly shows that life changes naturally. It does not require supernatural intervention. It does not require divine adjustment. It does not require a hidden designer.

The more we study life, the more we find mechanisms:

Mutation.

Inheritance.

Selection.

Drift.

Migration.

Recombination.

Isolation.

Adaptation.

Extinction.

God is not needed to make these processes work.

"Only a Theory" Misunderstands Science

One common objection is that evolution is "only a theory."

This phrase is misleading.

In everyday language, a theory may mean a guess. But in science, a theory is a well-supported explanatory framework. Germ theory explains infectious disease. Atomic theory explains matter. Cell theory explains the basic structure of living organisms. Evolutionary theory explains the diversity and relatedness of life.

Calling evolution a theory does not make it weak.

It places it among the great organizing explanations of science.

Evolution is supported by observation, experiment, fossils, anatomy, genetics, geography, and direct evidence of change in living populations. It is not a guess. It is not a religious belief. It is not an atheist slogan. It is a scientific explanation that works.

The phrase "only a theory" often reveals confusion about science.

A mature reader should not be misled by it.

The real question is not whether evolution is called a theory.

The real question is whether it explains the evidence.

It does.

Divine design does not.

What Real-Time Evolution Does Not Show

We should also be honest.

Real-time evolution does not show every event in the history of life. We do not watch a land mammal become a whale in a human lifetime. We do not watch dinosaurs become birds in a laboratory. We do not watch the entire history of life unfold before our eyes.

But that is not necessary.

Science often studies processes that operate across different scales.

We can observe erosion in a stream without watching the entire Grand Canyon form.

We can observe gravity in daily life without watching a galaxy form.

We can observe inheritance in families without watching the entire history of human ancestry unfold in one generation.

Observed evolution shows that the mechanism works.

Fossils show long-term change. DNA shows relatedness. Anatomy shows shared structure. Development shows deep biological connections. Geography shows patterns of distribution. Laboratory and field studies show change in action.

Together, these lines of evidence form a powerful case.

Observed evolution is one part of that case, but it is an important part because it brings evolution into the present.

It shows that evolution is not dead history.

It is ongoing life.

The Beauty of a Changing World

Some people feel threatened by the idea that life changes.

They want fixed categories. They want permanent kinds. They want humans set apart. They want nature to reflect a stable divine plan.

But a changing world is not less beautiful.

It is more alive.

A fixed creation would be like a statue gallery. Each species would stand where it had been placed, complete and unchanging.

Evolution gives us something richer:

A living river of change.

Forms arise, branch, adapt, struggle, and sometimes vanish. The world is not a collection of finished objects. It is a history still unfolding.

This truth can be emotionally difficult, but it is also magnificent.

Life is not frozen.

Life responds.

Life experiments without intention.

Life preserves what works.

Life loses what fails.

Life carries its past into its future.

That is a deeper kind of wonder than instant creation.

It is wonder without worship.

The Calm Conclusion

Evolution in real time gives us a direct view of life changing.

We see it in antibiotic resistance.

We see it in pesticide resistance.

We see it in viruses.

We see it in laboratory experiments.

We see it in wild populations.

We see it in plants adapting to new pressures.

We see it whenever inherited variation meets selection across generations.

This does not explain every detail of life's history by itself, but it confirms the reality of the process. Evolution is not merely an old story buried in rock. It is a living process still at work.

This matters for the claim of divine design.

If life were fixed by God in separate categories, we would not expect constant natural change. If adaptation required divine planning, we would not expect blind selection to explain so much. If species were permanent products, we would not expect populations to shift, split, resist, adapt, and diversify.

But they do.

The living world behaves as evolution predicts.

It does not behave as special creation would lead us to expect.

The conclusion is clear:

Evolution is real, observable, and powerful.

Life changes naturally.

God is not required.

Chapter Summary

Evolution is not only ancient history. It can be observed in real time.

Evolution means change in inherited characteristics of populations over generations.

Antibiotic resistance is one of the clearest medical examples of natural selection.

Pesticide resistance shows the same process in agriculture.

Viruses can evolve rapidly because they replicate and change quickly.

Laboratory evolution allows scientists to study evolutionary change under controlled conditions.

Wild populations, including birds, moths, and small mammals, show how environmental pressures can shift traits.

Plants can evolve resistance, tolerance, and timing changes under selection pressure.

Artificial selection demonstrates how selection can reshape living populations.

The distinction between microevolution and macroevolution is often misused. Large evolutionary change is connected to accumulated smaller changes over long periods.

Observed evolution weakens the claim that life is fixed by divine creation.

Key Takeaway

Evolution is not a guess about the distant past. It is a process we can observe in the present. Bacteria become resistant, viruses change, insects adapt, plants shift, and animal populations respond to environmental pressure. These changes do not require divine guidance. They show that life is dynamic, natural, and continuously shaped by inherited variation, selection, and time.

Sources

National Academies of Sciences, Engineering, and Medicine. Science, Evolution, and Creationism.

University of California Museum of Paleontology. Understanding Evolution: Evolution in Action.

University of California Museum of Paleontology. Understanding Evolution: Natural Selection.

Centers for Disease Control and Prevention. About Antimicrobial Resistance.

Centers for Disease Control and Prevention. Antimicrobial Resistance Facts and Stats.

World Health Organization. Global Antimicrobial Resistance Surveillance Report 2025.

Richard E. Lenski and colleagues. Long-Term Experimental Evolution in Escherichia coli.

Peter R. Grant and B. Rosemary Grant. Studies of Darwin's Finches.

Jerry A. Coyne. Why Evolution Is True.

Carl Zimmer. Evolution: The Triumph of an Idea.

Transition to the Next Chapter

Evolution can be observed in the present, but the full story of life requires a much larger stage.

That stage is deep time.

Small changes can be measured today. Across millions and billions of years, small changes can accumulate into enormous transformations. Fossils preserve parts of that long story in stone.

The next chapter will take us into deep time and the fossil record, where the history of life appears not as sudden divine creation, but as gradual change, branching, extinction, and transformation.

Chapter 7: Deep Time and the Fossil Record

The Stone That Remembered Life

A child picks up a stone and notices a strange pattern on its surface.

It is not a drawing. It is not a scratch made by a human hand. It is the outline of something that once lived. Perhaps a shell. Perhaps a leaf. Perhaps the delicate trace of a creature that moved through ancient mud long before any human being existed to see it.

The child turns the stone in his hand and asks a simple question:

How old is this?

That question opens one of the deepest doors in science.

A fossil is not just an old object. It is a message from a vanished world. It tells us that life existed before us, changed before us, struggled before us, and disappeared before us. It reminds us that human beings arrived very late in the history of Earth.

For many ancient religions, the Earth was imagined as young, human-centered, and created in a short act of divine will. Life appeared as separate forms, placed into the world by God. Human beings stood near the center of meaning.

The fossil record tells a very different story.

It tells us that the Earth is ancient.

Life is ancient.

Species are not fixed.

Most forms of life are gone.

The living world has changed across immense time.

This is not the pattern of sudden perfect creation.

It is the pattern of natural history.

Why Human Time Misleads Us

Human beings think in short time.

We remember childhood. We think in years, decades, and generations. A hundred years feels long. A thousand years feels ancient. Ten thousand years feels almost beyond imagination.

But evolution requires a much larger scale.

It requires deep time.

The Earth is about 4.5 billion years old. Life has existed for billions of years. Complex multicellular life appeared much later. Dinosaurs lived and vanished long before human beings existed. Modern humans are very recent in the full history of life.

This is difficult for the mind to grasp.

If the entire history of Earth were compressed into one calendar year, human civilization would appear only in the final moments of December 31. The pyramids, the Vedas, the Bible, the Quran, the Buddha, the Roman Empire, modern science, hospitals, airplanes, computers, and all sacred books would occupy only the last seconds of that imaginary year.

That is humbling.

It also weakens the religious idea that the Earth was built around human beings.

For most of Earth's history, there were no humans. For most of life's history, there were no mammals. For vast stretches of time, there were no birds, no flowers, no primates, no human language, no prayer, no temple, no scripture, no priest, and no prophet.

Life existed without us for billions of years.

The Earth was not waiting quietly for human religion to arrive.

Deep Time Makes Evolution Understandable

Many people reject evolution because they imagine it happening too quickly.

They ask:

How could an eye evolve?

How could a wing evolve?

How could fish give rise to land animals?

How could land mammals give rise to whales?

How could small changes produce such major differences?

These questions seem difficult when we think only in human time.

Deep time changes the picture.

A single drop of water seems weak, but water over long ages can shape rock. A small stream can cut a valley. Wind can wear stone. Sediment can build layers. Pressure can transform mud into rock. Tiny changes, repeated long enough, become powerful.

The same is true in biology.

A small inherited difference in one generation may appear minor. But if similar small changes accumulate over thousands, millions, or tens of millions of years, lineages can change dramatically.

Evolution does not require a modern bird to appear suddenly from a reptile. It does not require a whale to appear suddenly from a land mammal. It does not require a human being to appear suddenly from an ape-like ancestor.

It requires populations to vary, reproduce, become separated, adapt, and accumulate differences over enormous time.

Deep time gives evolution room to work.

Without deep time, creation feels necessary.

With deep time, natural history becomes enough.

Rocks as Pages of the Past

The fossil record is written mainly in rock.

Sediments settle in oceans, lakes, rivers, floodplains, deserts, and ancient shorelines. Over long periods, layers accumulate. Under pressure, these layers can become rock. Sometimes the remains or traces of living organisms are buried within them. Under the right conditions, bones, shells, teeth, leaves, footprints, burrows, pollen, or other traces are preserved.

These fossils are not scattered randomly through Earth's history.

They appear in patterns.

Older rocks contain older forms of life. Younger rocks contain more recent forms. Certain organisms appear, flourish, and disappear. Some forms are found only in certain layers. Others persist for long periods. New groups appear after earlier groups. The order is not perfect in every local place because geology is complex, but the broad pattern is clear.

Life has changed through time.

This is not what we would expect if all species were created together in one short act. If every kind of life had been separately created at the same time, we might expect fossils to be mixed without deep historical order. We might find modern humans with trilobites, rabbits with early Cambrian marine life, flowering plants in the oldest fossil layers, or whales before land mammals.

We do not.

The fossil record shows sequence.

That sequence is evidence.

Fossils and the End of Fixed Species

The old creation view imagined species as fixed and separate.

But fossils show that the living world was not always as it is now.

There were once trilobites in ancient seas. They are gone. There were giant marine reptiles. They are gone. There were ammonites. They are gone. There were non-avian dinosaurs that dominated land environments for vast periods. They are gone. There were many human relatives. They are gone. Only Homo sapiens remains.

If species were fixed creations, why does the history of life contain so many vanished forms?

Why do older layers contain organisms very different from those alive today?

Why do new forms appear after earlier forms?

Why do entire groups disappear?

Why do fossils often show features linking major groups?

Creation can always answer, "God made it that way."

But that is not an explanation. It does not explain the pattern. It does not predict the sequence. It does not show why certain fossils appear where they do. It does not explain why extinct forms often resemble earlier or later life in ways that fit evolutionary relationship.

Evolution explains this naturally.

Species appear because populations change and split.

Species disappear because extinction is common.

Fossils appear in sequence because life has a history.

Similarities exist because of ancestry.

The fossil record does not look like a display room of separate divine products.

It looks like a family history with many dead branches.

Transitional Fossils: The Missing Links That Were Found

Religious critics often ask, "Where are the missing links?"

This question is usually asked as if science has found none.

That is false.

Many transitional fossils have been found.

A transitional fossil does not need to be a perfect half-and-half creature. That is a cartoon version of evolution. A transitional fossil is a form that shows features connecting earlier and later groups. It may have a mixture of traits. It may help show how one kind of body plan changed over time.

Archaeopteryx is one famous example. It had feathers, but also teeth, a long bony tail, and clawed fingers. It helped show connections between dinosaurs and birds.

Tiktaalik is another important example. It was a fish-like animal with features that help illuminate the transition toward four-limbed vertebrates. It had fish features, but also a neck, strong fins with bones comparable in important ways to limb bones, and other traits useful for life in shallow water.

Whale evolution also has a rich fossil record. Early whale relatives show a transition from land-dwelling mammals to fully aquatic whales. Some had legs. Some could move on land and in water. Later forms became more fully adapted to aquatic life. The nostrils shifted position over time toward the blowhole. Limbs changed. Bodies became more streamlined.

These examples do not show sudden creation.

They show history.

The fossil record is incomplete, but it is not empty. The "missing link" argument has become weaker because many predicted links have been found.

And the incompleteness of the fossil record is not surprising.

Fossilization is rare. Most organisms die and decay without becoming fossils. Soft bodies preserve poorly. Many fossils are destroyed by erosion, heat, pressure, or geological change. Many remain buried and undiscovered. We should not expect the fossil record to contain every organism that ever lived.

But the fossils we do have are enough to show a powerful pattern of change.

Fish to Land Animals

One of the great transitions in the history of life was the movement of vertebrates from water to land.

This did not happen in one magical leap.

The ancestors of land vertebrates were fish-like animals. Over time, some lived in shallow water environments where certain traits could be useful: stronger fins, air breathing, neck mobility, and the ability to push through vegetation, mud, or shallow water.

Fossils show forms that help connect fish to early four-limbed animals. These fossils are not modern fish and not modern amphibians. They represent stages in an evolutionary story.

This matters because creation thinking often imagines sharp categories.

Fish are fish.

Land animals are land animals.

Each kind is separate.

But nature does not always respect our categories. Evolution works through populations, variation, and gradual change. A creature can have fish-like features and also features that foreshadow land vertebrates.

The boundary is not a wall.

It is a transition.

The religious view of separate creation struggles with such forms. Why create animals that look like historical bridges? Why place them in the fossil record in the expected sequence? Why give them combinations of traits that fit evolutionary prediction?

Evolution explains them naturally.

They are not divine puzzles.

They are ancestral clues.

Land Mammals to Whales

Whales are among the most striking examples of evolutionary transformation.

A whale looks nothing like a deer, wolf, or hippo to the casual observer. It lives in water, has flippers, a streamlined body, and breathes through a blowhole. It seems like a creature made for the sea.

But fossils, anatomy, embryology, and DNA reveal a deeper story.

Whales are mammals.

They breathe air.

They nurse their young.

They have remnants of pelvic bones.

Their embryos show developmental clues.

Their DNA connects them to other mammals.

Fossils show early whale relatives that were more land-based, then more amphibious, then more fully aquatic.

This is not what separate creation would naturally suggest.

If whales were simply created as ocean creatures, why do they carry mammalian features? Why do they breathe air instead of having gills? Why do they nurse their young? Why do they show anatomical traces of land ancestry? Why does their fossil history show a transition from land to water?

Evolution makes sense of these facts.

Whales were not designed from nothing as perfect sea machines. They are descendants of land mammals whose lineage returned to the water over millions of years.

Their bodies are not evidence of sudden divine creation.

They are evidence of transformation.

Dinosaurs and Birds

Birds are living beauty.

They sing, migrate, nest, fly, and fill the sky with movement. For many people, birds feel like symbols of freedom, spirit, and divine artistry.

But birds are also living dinosaurs.

Modern birds are descendants of theropod dinosaurs. This conclusion is supported by fossils, anatomy, feathers, bone structure, eggs, nesting behavior, and many other lines of evidence.

Feathered dinosaurs have been discovered. Some had feathers not necessarily used for powered flight. Feathers may have served earlier functions such as insulation, display, or brooding before they became central to flight in certain lineages.

This is important because it shows how evolution can modify a structure over time.

A feature may begin with one function and later serve another.

A divine designer could create birds instantly with wings and feathers for flight. But the evidence shows a deeper history. It shows earlier feathered forms, gradual changes, and connections between birds and dinosaurs.

The sky is not less beautiful because birds evolved.

It is more meaningful.

Every bird in flight carries the memory of ancient life.

The Horse, Carefully Understood

Horse evolution has often been presented as a straight line from a small ancient animal to the modern horse.

That picture is too simple.

Horse evolution was not a neat ladder. It was a branching history with many species, side branches, and extinctions. Some forms had different numbers of toes. Some were adapted to forests. Later forms became better suited to open grasslands. Teeth, limbs, and body size changed across different lineages.

This is important because evolution should not be explained as a simple march toward a modern goal.

The modern horse was not the pre-planned destination of ancient horse-like animals. It is one surviving branch of a larger and more complex history.

This point matters for the whole book.

Evolution is not a ladder.

Evolution is not progress toward perfection.

Evolution is not a divine staircase.

Evolution is branching, contingent, and full of extinction.

The horse does not show a straight road designed in advance. It shows changing conditions, branching lineages, and the survival of some forms while others disappeared.

That is natural history.

Not divine planning.

Human Fossils: A Preview

Human evolution will be discussed more fully in a later chapter, but the fossil record already gives us a preview.

Humans did not appear suddenly as separate creations.

The fossil record shows many forms related to our ancestry and to our broader human family. Australopithecines, early Homo species, Neanderthals, Denisovans, and others reveal that human history is not a simple story of one created pair. It is a branching history of populations, migrations, adaptations, interbreeding, and extinctions.

This is emotionally difficult for many religious readers.

The idea of special creation gives humans a sacred beginning. The fossil record gives us a natural beginning. It places us among other primates, not above nature as separate divine objects.

But this truth need not degrade us.

The fossil record does not make humans worthless. It makes us historical. It shows that our bodies, brains, hands, gait, and social life developed through a long natural process.

We are not less valuable because we have ancestors.

We are more deeply connected.

The Fossil Record Is Incomplete, But Not Weak

Creationists often point to gaps in the fossil record.

There are gaps. No honest scientist denies this.

But gaps are expected.

Fossilization is rare. Most living things do not fossilize. Many organisms have soft bodies that decay quickly. Many fossils are destroyed by geological processes. Many remain hidden under oceans, mountains, forests, cities, deserts, and ice. The deeper we go into time, the more difficult preservation becomes.

The question is not whether the fossil record contains every life form that ever existed.

It does not.

The real question is whether the fossil record we have supports evolution or separate creation.

It supports evolution.

Fossils appear in meaningful sequence. Transitional forms exist. Extinct organisms often fit expected historical patterns. Older rocks contain older forms. Major groups do not appear randomly mixed throughout all layers. Fossils are found in the kinds of places and ages where evolutionary theory would lead us to search.

A partial record can still be powerful.

A torn family album may not contain every photograph, but it can still show childhood, youth, adulthood, relationships, and ancestry. A damaged history book may have missing pages, but it can still reveal the direction of events.

The fossil record is incomplete, but it speaks clearly.

Life has changed.

Radiometric Dating and the Age of Rocks

How do scientists know the age of rocks and fossils?

One important method is radiometric dating.

Certain atoms are unstable and decay into other atoms at known rates. By measuring the ratio of parent and daughter isotopes in rocks or minerals, scientists can estimate when those minerals formed. Different dating methods apply to different materials and time ranges.

This is not guesswork based on personal opinion.

Radiometric dating is grounded in physics. It can be tested, compared, and refined. Different methods can often be used together. Dates can be checked against geological layers, volcanic ash, magnetic reversals, and other evidence.

Fossils themselves are not always dated directly. Often, scientists date the rock layers above or below them, especially volcanic layers, to estimate the age of fossil-bearing sediment.

The result is a coherent timescale for Earth's history.

This timescale is not compatible with a young Earth created only a few thousand years ago. It reveals vast ages before humans, before mammals, before dinosaurs, before land animals, and before complex life.

The evidence of deep time is one of the great blows against literal creation stories.

It shows that ancient scripture was not written as geology.

It was written by ancient human beings who did not know Earth's real age.

Why a Young Earth Fails

Some religious believers still defend a young Earth. They argue that the world is only thousands of years old and that fossils were created during a flood or placed in the ground by divine action.

This view fails scientifically.

It fails because evidence from geology, physics, astronomy, biology, and fossils all points to an ancient Earth. Rock layers show long histories of deposition, erosion, uplift, volcanic activity, climate change, and life. Coral reefs, ice cores, lake sediments, radiometric dating, plate tectonics, and starlight all point far beyond a few thousand years.

A young Earth view must reject too much evidence.

It must force nature to fit scripture instead of allowing evidence to speak for itself.

This is not science.

It is religious commitment protecting itself from reality.

A mature scientific view does not begin with a sacred date and then bend the Earth to match it. It begins with evidence.

And the evidence says Earth is ancient.

Once Earth is ancient, evolution becomes not only possible, but expected.

Fossils Do Not Show Human Centrality

The fossil record is deeply humbling because it does not place humans at the center.

For most of life's history, humans did not exist. For vast stretches of time, no creature prayed, wrote scripture, built temples, or asked theological questions. Life flourished and died without human witnesses.

If the world was designed mainly for humans, the fossil record is strange.

Why create billions of years of life before humans?

Why create entire worlds and erase them?

Why fill the planet with creatures that lived, suffered, and vanished long before any human moral drama?

Why make humans arrive so late?

Evolution answers naturally.

Humans arrived late because we evolved late. We are not the purpose of the whole story. We are one recent branch of a much older tree.

This conclusion may offend human pride, but it fits the evidence.

The fossil record does not tell a human-centered religious story.

It tells a natural story.

Extinction in the Rocks

Fossils do not only show life.

They show death.

They show that species and entire groups can disappear forever. Trilobites are gone. Ammonites are gone. The non-avian dinosaurs are gone. Many ancient marine reptiles are gone. Many early human relatives are gone. Countless known and unknown species have vanished.

This is not a minor detail.

Extinction is one of the strongest arguments against perfect divine design.

A designed world might be expected to show purpose, care, and preservation. But the fossil record shows waste on a scale the human mind can hardly hold. Entire ecosystems rise and fall. Beautiful and complex organisms vanish. Mass extinctions erase huge portions of life.

The religious answer may be that God had a mysterious reason.

But mystery is not explanation.

Evolutionary biology, geology, ecology, climate science, and planetary science offer real explanations: environmental change, competition, disease, volcanic eruptions, asteroid impact, sea-level changes, oxygen changes, and ecological collapse.

The fossil record does not need divine mystery.

It needs natural history.

The Pattern Creation Cannot Explain Well

The fossil record creates a serious problem for separate creation.

If all life were created separately by God, why does the fossil record show an order that fits evolution?

Why are simpler marine forms found in very old rocks?

Why do land vertebrates appear after aquatic life?

Why do mammals become more prominent after the extinction of non-avian dinosaurs?

Why do humans appear extremely late?

Why are transitional fossils found in predicted geological contexts?

Why do extinct species often show mixtures of traits?

Why are modern animals not found randomly throughout all fossil layers?

Creation can only answer, "God chose it that way."

But that answer explains nothing. It gives no mechanism. It makes no testable prediction. It does not help scientists know where to search, what to expect, or how to connect evidence.

Evolution does.

Evolution predicts that life should appear in historical sequence. It predicts that transitional forms should be found in certain layers and not others. It predicts that older ancestors should appear before later descendants. It predicts branching, extinction, and anatomical continuity.

The fossil record fits that pattern.

That is why fossils matter.

They are not merely old bones.

They are evidence against sudden divine creation.

The Beauty of a Non-Miraculous History

Some readers may feel sadness when they first accept the deep history of life.

The old story was simple.

God made life.

God made humans.

Humans were the center.

The world was spiritually arranged.

The fossil record tells a less comforting but more magnificent story.

Life began in ancient seas. It changed slowly. It spread. It adapted. It suffered. It survived catastrophes. It produced countless forms. Most vanished. Some left descendants. A branch gave rise to vertebrates. Some vertebrates moved onto land. Some lineages returned to water. Some took to the air. Mammals diversified. Primates evolved. Human ancestors walked upright. Minds grew. Language appeared. Culture emerged. Science was born. Finally, a species that evolved from nature became capable of understanding nature.

That is not a small story.

It is larger than myth.

It does not need miracles because reality itself is profound.

The fossil record does not make life less wonderful. It makes life older, stranger, deeper, and more connected.

A god did not need to place every creature separately into the world.

Nature had time.

And time changed everything.

Deep Time and Science-Based Atheism

Deep time is essential to science-based atheism because it removes the need for sudden creation.

A young Earth needs miracles.

An ancient Earth permits natural history.

If life had only a few thousand years, evolution could not explain much. But with billions of years of Earth history, millions of generations, shifting environments, extinction, isolation, mutation, selection, and inheritance, the living world becomes understandable without divine design.

This is why many religious traditions resisted deep time. It did not merely change the age of the Earth. It changed the meaning of life.

It showed that humans were late arrivals.

It showed that scripture did not contain scientific history.

It showed that life changed before humans existed.

It showed that nature had its own story independent of religion.

Once deep time is accepted, the old creation picture becomes much harder to defend.

The Earth is not a young stage built for human beings.

It is an ancient planet with a long, natural, and often brutal history.

The Calm Conclusion

The fossil record does not tell us that every question has been answered. Science never needs to pretend that. There are gaps. There are debates. There are new discoveries waiting underground. There are details still being refined.

But the broad message is unmistakable.

Life has changed through time.

Species are not fixed.

Extinction is real and massive.

Transitional forms exist.

Humans appeared late.

The Earth is ancient.

The fossil record fits evolution far better than separate creation.

The old religious story said life was created in fixed forms.

The rocks say life has a history.

The old religious story placed humans at the center.

The rocks say humans arrived near the end of a very long biological story.

The old religious story saw divine planning.

The rocks show change, branching, catastrophe, survival, and extinction.

This does not make life meaningless.

It makes life honest.

The fossil record is not a hymn to special creation. It is a vast archive of natural history. It shows that life does not need to be explained by a supernatural designer.

It has been shaped by time, change, survival, and death.

No divine creation is required.

Chapter Summary

Human beings naturally think in short time, but evolution requires deep time.

The Earth is about 4.5 billion years old, and life has existed for billions of years.

Fossils appear in meaningful sequence, not as a random mixture of all species.

The fossil record shows that species are not fixed and that many forms of life have vanished.

Transitional fossils, such as Archaeopteryx, Tiktaalik, and early whale relatives, support evolutionary change.

The fossil record is incomplete, but its broad pattern strongly supports evolution.

Radiometric dating and geology reveal an ancient Earth, not a young creation.

Extinction is a major problem for the idea of perfect divine design.

Humans appear very late in the history of life, which weakens human-centered creation stories.

Deep time removes the need for sudden divine creation and allows life to be understood through natural history.

Key Takeaway

The fossil record shows that life was not created suddenly in fixed forms. It changed across deep time. Species appeared, adapted, branched, and vanished. Transitional fossils, extinction, geological layers, and the late arrival of humans all point toward natural history, not divine creation. The rocks do not reveal a perfect designer. They reveal evolution written in stone.

Sources

Charles Darwin. On the Origin of Species.

Charles Lyell. Principles of Geology.

National Academies of Sciences, Engineering, and Medicine. Science, Evolution, and Creationism.

National Academies of Sciences, Engineering, and Medicine. Teaching About Evolution and the Nature of Science.

University of California Museum of Paleontology. Understanding Evolution: Fossil Evidence.

University of California Museum of Paleontology. Understanding Evolution: Transitional Features.

United States Geological Survey. Geologic Time and Radiometric Dating Resources.

Donald R. Prothero. Evolution: What the Fossils Say and Why It Matters.

Neil Shubin. Your Inner Fish.

Jerry A. Coyne. Why Evolution Is True.

Carl Zimmer. Evolution: The Triumph of an Idea.

Transition to the Next Chapter

Deep time and fossils show that life has changed across Earth's long history.

But another question remains.

How did life begin in the first place?

Evolution explains how living beings change after life exists. The origin of life asks how chemistry became biology. Religious belief often hides God inside that question.

The next chapter will examine the origin of life carefully and honestly, without pretending that science knows everything and without pretending that every unknown is evidence for God.

Chapter 8: Origin of Life: Chemistry Before Biology

The Question Religion Wants to Keep

There is one question religious belief loves to keep for itself:

How did life begin?

For many believers, this question feels like the final safe room for God. They may accept some fossils. They may accept that animals change a little over time. They may even accept that natural selection can explain some adaptation. But then they stop and ask:

Where did the first life come from?

This question is powerful because it reaches into one of the deepest mysteries of science. Before the first cell, there was no bird, no tree, no fish, no eye, no heart, no brain, and no organism like anything alive today.

There was chemistry.

There was water, rock, heat, minerals, lightning, volcanic activity, sunlight, radiation, gases, tides, impacts, and energy. There were atoms and molecules interacting under conditions very different from the world we know now.

At some point in Earth's deep past, nonliving chemistry crossed into something life-like.

That transition is one of the most profound questions human beings can ask.

It is also one of the most misused questions in religion.

The religious argument usually moves too quickly:

Science cannot fully explain the first life.

Therefore, God created life.

But that conclusion does not follow.

A mystery is not evidence of God.

A gap is not proof of a miracle.

An unanswered question is not a divine signature.

The honest scientific answer is more careful:

We do not yet know every step by which life began, but there are natural pathways worth studying.

That answer may sound less dramatic than "God did it."

But it is more honest.

Evolution and the Origin of Life Are Different Questions

Before we go further, one confusion must be removed.

Evolution and the origin of life are related, but they are not the same question.

Evolution explains how living populations change after life already exists. It explains variation, inheritance, natural selection, adaptation, speciation, common ancestry, extinction, and the diversity of life.

The origin of life asks a different question:

How did the first self-sustaining, self-replicating, evolving chemical systems arise from nonliving matter?

This field is often called abiogenesis.

Religious critics often confuse these two subjects. They say, "Evolution cannot explain how life began." But that is like saying cardiology cannot explain how the first hospital was built. It is not the same question.

Evolution is about the history of life after life begins.

Origin-of-life science is about how chemistry may have become biology.

This distinction matters because some critics attack evolution by demanding that it explain the first living system. But even if every detail of the origin of life were still unknown, evolution would remain strongly supported by fossils, DNA, anatomy, observed evolution, and common ancestry.

Still, the origin of life is a real scientific question.

Science does not run away from it.

Science studies it.

The Early Earth Was Not Empty

The early Earth was not a quiet garden waiting for divine breath.

It was a young, active planet. Volcanoes erupted. Asteroids and comets struck. Oceans formed. Heat moved through rock. Lightning flashed. Minerals reacted with water. Chemicals were exposed to energy, wet-dry cycles, temperature changes, pressure, radiation, and time.

This was not life yet.

But it was not nothing.

It was chemistry.

The central idea of origin-of-life research is that under certain conditions, simple molecules can form more complex molecules. Some molecules can organize. Some can help reactions occur. Some can form compartments. Some may store information. Some chemical systems may begin to copy parts of themselves imperfectly. Once imperfect copying and variation exist, a primitive form of selection can begin.

This is not magic.

It is chemistry moving toward biology.

No one should pretend that every step is fully known. It is not. But ignorance of every detail does not justify inserting God. There was a time when human beings did not understand lightning, infection, heredity, mental illness, or the motion of planets. The lack of knowledge did not make supernatural explanations true.

The early Earth was a chemical world.

The question is not whether God can be imagined there.

The question is whether natural processes can be investigated there.

They can.

The Building Blocks Are Natural

Modern life depends on complex molecules: proteins, nucleic acids, lipids, carbohydrates, and many others. But these did not need to appear all at once in their modern form.

Origin-of-life research asks how simpler chemical building blocks could form under early Earth conditions or arrive from space.

Amino acids, which are building blocks of proteins, can form under certain nonliving chemical conditions. Organic molecules have been found in meteorites and asteroid samples. Nucleobases, which are part of RNA and DNA chemistry, can arise in nonliving contexts. Lipid-like molecules can form simple membrane-like structures in water.

These facts do not prove exactly how life began.

But they prove something important:

The ingredients of life are not beyond nature.

They are chemical.

This weakens the religious claim that life's building blocks require direct divine manufacture. Science has repeatedly shown that molecules once considered special to life can arise through natural chemistry.

The old idea of a mysterious life-force has faded. Living beings are not made of supernatural material. They are made of atoms and molecules: carbon, hydrogen, oxygen, nitrogen, phosphorus, sulfur, metals, salts, water, and many other chemical components.

The difference between life and non-life is not that one contains magic and the other does not.

The difference is organization, information, replication, metabolism, compartments, and evolution.

Molecules From Space

Some of life's building blocks may also have come from space.

Meteorites and asteroids can contain organic molecules. Asteroid sample studies have strengthened the evidence that life-related molecules can form in nonliving environments beyond Earth.

This does not mean life itself came from space.

It does not prove that biology began elsewhere.

It does not mean asteroids carried living cells to Earth.

But it does show that important chemical ingredients can arise naturally in the wider solar system and may have been delivered to the early Earth.

This is a powerful point.

It means prebiotic chemistry is not restricted to one small pond, one ocean vent, or one location on Earth. The universe is chemically active. Organic chemistry can occur in space, inside asteroids, in meteorites, and in planetary environments.

A religious argument may say, “Life’s molecules are too special. God must have made them.”

Science replies, “Let us look.”

And when science looks, it finds organic chemistry in natural settings.

The universe is not dead matter waiting for a miracle.

It is a place where chemistry can become surprisingly rich.

The RNA World

One major idea in origin-of-life research is the RNA world hypothesis.

Today, life uses DNA to store genetic information, proteins to perform many cellular tasks, and RNA as a crucial messenger and worker molecule. But this creates a puzzle. DNA depends on proteins for copying and maintenance. Proteins are made using genetic instructions. So which came first?

RNA offers a possible bridge.

RNA can store information, somewhat like DNA. RNA can also perform catalytic functions, somewhat like proteins. This makes it a possible early molecule that could carry information and help chemical reactions occur.

In the RNA world idea, early life-like systems may have used RNA before the modern DNA-protein system evolved.

This does not mean scientists have solved everything.

Important questions remain.

How were the first RNA building blocks formed?

How did they join into longer chains?

How did copying begin?

How accurate did early copying need to be?

How were RNA-like systems protected from destruction?

How did RNA systems become enclosed in compartments?

How did DNA and proteins later take over many modern roles?

These are serious questions.

But serious questions are not defeats for science. They are the normal work of science.

The RNA world hypothesis is valuable not because it answers everything, but because it provides a testable path. Scientists can study RNA chemistry, catalytic RNA, nucleotide formation, polymerization, early compartments, and conditions that could support such processes.

"God did it" does none of this.

It does not tell us what to test.

It does not predict what molecules should form.

It does not explain how heredity began.

It does not explain how chemistry crossed into biology.

It does not guide laboratory research.

Science keeps going.

Protocells: The First Boundaries

Modern cells have membranes.

A membrane separates inside from outside. It allows a cell to maintain a special internal environment. It helps organize chemistry. It keeps useful molecules near one another.

A key question in origin-of-life research is how early compartments could have formed.

Simple lipid-like molecules can sometimes organize themselves into membrane-like structures in water. This happens because of their chemical properties. Some parts of the molecule interact well with water. Other parts avoid water. Under the right conditions, these molecules can form bubbles or vesicles.

These are not modern cells.

But they may help us imagine early protocells.

A protocell would not need to be alive in the full modern sense. It might be a simple compartment containing molecules that interact, grow, divide, or preserve certain chemical systems better than others. If some protocells were more stable, better at capturing useful molecules, or better at allowing replication inside them, those systems could persist more successfully.

This matters because life needs organization.

Chemistry floating freely in an ocean may not easily become biology. Compartments concentrate molecules. They create inside and outside. They allow certain chemical combinations to remain together long enough to matter.

Again, no miracle is required to study this.

The question is chemical and physical:

How can molecules organize into compartments?

That question can be tested.

Energy: Life Needs More Than Ingredients

A pile of ingredients is not life.

Modern cells need energy. They do not merely contain chemicals. They manage energy flow. They use gradients. They build molecules, repair damage, move materials, copy information, and maintain order.

So origin-of-life science must ask not only where life's building blocks came from, but how early chemical systems were powered.

This is one reason hydrothermal vents, hot springs, mineral surfaces, volcanic environments, and wet-dry cycles are studied. These settings can provide energy differences, chemical gradients, heat, minerals, and repeated cycles that may help drive reactions.

A recipe is not enough.

There must be a kitchen.

There must be conditions that allow ingredients to interact in useful ways.

This makes the origin of life a difficult problem, but not a supernatural one. It is a problem of chemistry, physics, geology, energy, and time.

Religion says, “God breathed life into matter.”

Science asks:

What molecules were present?

What energy sources existed?

What environments concentrated the molecules?

What minerals helped reactions?

What cycles built complexity?

What systems could copy with variation?

Those questions lead to investigation.

That is why they are stronger than a miracle story.

Hydrothermal Vents: Chemistry in the Deep

One important idea places the origin of life near hydrothermal vents.

Hydrothermal vents are places where heated fluids emerge through rock, often in ocean settings. These environments can provide minerals, heat, chemical gradients, and energy sources. Some scientists think such settings may have helped early chemistry become more organized.

The idea is attractive because life depends on energy gradients. Modern cells use controlled chemical differences to power work. At vents, natural gradients exist between hot fluids and surrounding water, between different chemicals, and between mineral surfaces and fluid environments.

Mineral surfaces may help concentrate molecules or catalyze reactions. Tiny pores in rocks may act like natural compartments. Energy gradients may help drive chemical processes.

This does not mean hydrothermal vents are the final answer.

Science has not settled the matter.

Other environments may also have been important. Shallow ponds, hot springs, tidal pools, volcanic fields, ice, mineral surfaces, and wet-dry cycles have all been considered.

The important point is that scientists are not staring helplessly into darkness.

They are comparing natural environments. They are asking which conditions could produce building blocks, organize them, energize them, protect them, concentrate them, and allow them to develop toward life-like systems.

Religion offers one sentence: God created life.

Science offers many testable questions.

That is why science moves forward.

Warm Little Ponds and Wet-Dry Cycles

Darwin once imagined that life might have begun in some "warm little pond," with chemicals, light, heat, and energy interacting under favorable conditions.

Modern research has given serious attention to land-based settings such as hot springs, shallow ponds, and volcanic pools. One reason is that wet-dry cycles can help chemical reactions that are difficult in open water. When water evaporates, molecules can become concentrated. When water returns, mixing and further reaction can occur. Repeated cycles may help build more complex molecules.

This is not a simple answer.

The early Earth had many environments. The origin of life may not have occurred in one neat location. It may have required multiple settings, repeated chemical pathways, and many failed attempts before a stable evolving system appeared.

But the larger point is clear:

Scientists can imagine and test natural environments where chemistry becomes more complex.

This is very different from saying, "We do not know, therefore God."

That phrase has failed too many times in human history. The safer and more honest path is to investigate.

The origin of life may have been rare. It may have required unusual conditions. It may have involved a long sequence of chemical steps.

But rarity is not miracle.

Difficulty is not impossibility.

The First Replication

Life as we know it depends on reproduction. But before full cells existed, there may have been simpler forms of copying.

The first self-replicating systems may not have looked anything like modern organisms. They may have been chemical systems capable of making rough copies of themselves or helping related molecules form.

These copies would not need to be perfect.

In fact, perfect copying would stop evolution from having raw material. Imperfect copying creates variation.

Once a system can copy itself with variation, and once some variations persist better than others, a primitive form of evolution can begin.

This is one of the most important bridges between chemistry and biology.

Replication creates continuity.

Variation creates possibility.

Selection creates direction without purpose.

The earliest replication may have been crude, unstable, and inefficient. But it did not need to begin as modern DNA replication. Nature does not require modern perfection at the first step. It requires a pathway from simpler systems to more stable ones.

Religious thinking often imagines that life must begin fully formed or not at all.

Science does not require that.

Science studies intermediate steps.

Metabolism First, Genes First, or Both Together?

Scientists do not all agree on the exact order of life's earliest features.

Some emphasize RNA-first scenarios, where information and replication came early. Some emphasize metabolism-first ideas, where networks of chemical reactions developed before genetic systems. Some emphasize compartments first. Others think early life required information, metabolism, and compartments to emerge together in interacting stages.

This disagreement is not weakness.

It is science doing its work.

In religion, disagreement often leads to sects, authority claims, and appeals to sacred texts. In science, disagreement leads to experiments, models, evidence, correction, and better questions.

Origin-of-life research is difficult because the earliest stages left limited direct evidence. The first life did not leave bones. It did not write messages in rock. The chemistry was tiny, ancient, and fragile. Much of the evidence must be reconstructed through chemistry, geology, biology, astronomy, and laboratory work.

So we should be honest:

Science does not yet possess a complete, final, universally accepted pathway from nonliving chemistry to the first living system.

But this honesty does not help religion as much as believers think.

A scientific unknown is not evidence for God.

It is evidence that a question remains open.

"We Do Not Know" Is Not "God Did It"

The most important sentence in this chapter may be this:

"We do not know yet" does not mean "God did it."

This mistake has been made again and again.

People did not know what caused lightning, so they imagined gods.

They did not know what caused disease, so they imagined spirits, curses, or divine punishment.

They did not know what caused mental illness, so they imagined demons.

They did not know how planets moved, so they imagined divine control.

They did not know how species adapted, so they imagined a designer.

In each case, supernatural explanations filled a gap.

Then science advanced.

The gap became smaller.

The god who lived inside that gap had to retreat.

This is why placing God inside the origin of life is risky for religion. It makes God dependent on ignorance. It says, in effect, "Here is something science has not fully explained yet. That is where God is."

But what happens if science explains more tomorrow?

A God who survives only in the shadows of current ignorance is not a strong God. Such a God becomes smaller every time knowledge grows.

Science-based atheism does not need to claim that every detail of the first life is already known.

It only needs to point out that no scientific evidence requires God as the explanation.

At present, origin-of-life research is an open scientific field.

It is not evidence for divine creation.

Why "God Did It" Explains Nothing

Suppose someone says, "God created the first life."

What has been explained?

How did God do it?

What physical process was used?

Why that chemistry and not another?

Why use RNA, DNA, proteins, membranes, and metabolism?

Why make life through chemistry at all?

Why make life so fragile?

Why create a world where life immediately enters struggle, mutation, competition, disease, and death?

Why wait billions of years before humans appear?

The statement "God created life" answers none of these questions scientifically.

It gives a conclusion without a mechanism.

A scientific explanation must do more. It must connect causes to effects. It must allow investigation. It must make contact with evidence. It must be open to testing and correction.

Origin-of-life science may be incomplete, but it moves in the right direction.

It asks what chemicals were present.

It asks what energy sources existed.

It asks how molecules formed.

It asks how they became concentrated.

It asks how compartments arose.

It asks how copying began.

It asks how selection entered chemistry.

It asks how early systems became more life-like.

This is explanation.

"God did it" is not.

It is a stop sign.

The First Life Was Not a Modern Cell

When people imagine the first life, they often imagine a complete modern cell.

That makes the origin of life seem impossible.

A modern cell is extraordinarily complex. It contains genetic information, proteins, membranes, ribosomes, energy systems, enzymes, transport systems, and many coordinated processes.

But the first life was almost certainly much simpler than modern life.

It may not have had DNA as modern cells do. It may not have used proteins in the modern way. It may not have had a full modern membrane. It may not have had the elegant machinery seen in bacteria today.

Modern cells are the result of billions of years of evolution.

It is a mistake to demand that the first life begin as a modern cell.

That is like looking at a modern city and saying the first human shelter must have had elevators, plumbing, electricity, hospitals, roads, and internet. It confuses the later result with the beginning.

The first life-like systems may have been crude.

But crude does not mean impossible.

A simple beginning can lead to complexity if there is copying, variation, selection, and time.

The Problem of Probability

Another religious argument says that life is too improbable to arise naturally.

This argument often sounds mathematical, but it is usually misleading.

It imagines modern life appearing all at once by pure chance. It calculates the probability of a complete modern protein, gene, cell, or organism forming randomly in a single step. Of course, that probability is extremely low.

But that is not what origin-of-life science proposes.

Science does not say a modern cell assembled instantly from random atoms.

It studies stepwise chemistry, environmental concentration, energy gradients, self-organization, catalysis, compartments, copying, selection, and gradual increases in complexity.

Probability changes when there are steps. It changes when chemistry limits what can happen. It changes when certain reactions are favored by physical conditions. It changes when intermediate products are preserved. It changes when selection begins.

A snowflake does not form because water molecules randomly arrange themselves into beauty with no rules. It forms because physical laws guide the structure.

Crystals form because chemistry and physics constrain possible arrangements. Biological chemistry also operates under constraints.

Chance matters.

But chance is not the whole story.

Natural law, chemistry, environment, energy, and selection matter too.

Chemistry Is Not Trying to Become Life

We must also avoid the opposite mistake.

Science should not speak as if chemistry wanted to become life.

Molecules have no ambition. RNA does not dream. Lipids do not plan. Minerals do not seek biology. The early Earth was not consciously moving toward cells.

There was no goal.

There was no foresight.

There was no divine script.

If life began naturally, it was not because the universe wanted humans. It was because matter under certain conditions can become organized in extraordinary ways.

This keeps the argument honest.

Science-based atheism should not replace God with a romantic view of nature. Nature is not a hidden mind. Nature is not a loving mother guiding molecules toward us. Nature has no intention.

That is the beauty and severity of the scientific view.

Life emerged, if science is correct, not because the universe planned us, but because matter and energy can produce complexity under the right conditions.

From Chemistry to Evolution

At some point, a crucial threshold was crossed.

A chemical system began to preserve information, copy itself imperfectly, and produce variations that affected persistence or reproduction.

Once that happened, evolution could begin.

This is where chemistry becomes biology.

Before that threshold, there is prebiotic chemistry. After it, there is the beginning of Darwinian evolution in its simplest form.

The first evolving systems were almost certainly very different from modern organisms. But once evolution began, natural selection could build complexity step by step. More stable systems persisted. Better copying improved inheritance. Useful chemical functions were preserved. Compartments became more effective. Metabolism became more organized. Later, DNA became a more stable information molecule. Proteins became powerful molecular tools. Cells became more sophisticated.

This was not instant creation.

It was a transition.

The religious mind often wants a sharp line: dead matter here, living soul there. But nature may not provide such a clean boundary. The origin of life may have been gradual, with systems becoming more life-like over time.

That idea may feel strange, but it is scientifically reasonable.

Many boundaries in nature are gradual.

Dawn is not night, but it is not full day.

A child becomes an adult gradually.

A language changes slowly.

A species can split over time.

Life may also have emerged through stages.

Why This Does Not Diminish Life

Some people feel that explaining life through chemistry makes it less special.

But why should that be true?

A rainbow is not less beautiful because we understand light. A flower is not less beautiful because we understand pollination. Love is not less real because the brain and body are involved. A newborn's first breath is not less moving because we understand lungs, oxygen, and physiology.

Understanding does not destroy wonder.

It deepens it.

The idea that chemistry could become life is not degrading. It is astonishing. It means matter is capable of organization, memory, adaptation, and eventually consciousness under the right conditions. It means the universe does not need magic to become marvelous.

A scientific view of life does not make life cheap.

It makes life rare and precious.

If life emerged naturally from the chemistry of an ancient planet, then every living being is part of a continuous story from atoms to cells, from cells to organisms, from organisms to minds, from minds to science, and from science to self-understanding.

That story is more moving than a simple command from heaven.

It is not smaller than religion.

It is larger.

The Religious Retreat Into the First Spark

As science explained more and more of life, God was pushed into earlier and smaller spaces.

When species seemed fixed, God created every species.

When evolution explained species, God guided evolution.

When natural selection explained adaptation, God became the source of the first life.

When prebiotic chemistry began to explain life's building blocks, God became the mysterious first spark behind chemistry itself.

This retreat is familiar.

God moves to the edge of knowledge.

But the edge of knowledge is not a stable home.

It keeps moving.

The first spark argument may sound strong today because the origin of life remains incompletely understood. But the pattern of history should make us cautious. Many things once thought impossible without God are now understood naturally.

There is no scientific reason to stop that expectation at the origin of life.

We should not pretend we know every detail.

But we should also not pretend that ignorance is evidence.

The Origin of Life and the Larger Argument

This chapter does not claim that science has solved the origin of life completely.

It has not.

But the chapter does claim something very important:

The origin of life does not provide scientific evidence for God.

There are natural research programs. There are plausible chemical pathways. There are known organic molecules. There are energy sources. There are models of protocells. There are hypotheses about RNA, metabolism, hydrothermal vents, hot springs, wet-dry cycles, mineral surfaces, and molecules from space.

These may not yet form a complete answer.

But they form a scientific path.

God is not a scientific path.

God does not explain the chemistry.

God does not explain the sequence.

God does not explain the mechanism.

God does not predict the evidence.

God does not guide experiment.

The origin of life remains a frontier of science, not a victory for religion.

The Calm Conclusion

The beginning of life is one of the deepest questions human beings can ask.

We should approach it with humility.

Science does not know everything. It should never pretend otherwise. The first steps from chemistry to biology remain difficult to reconstruct. The evidence is ancient, fragile, and incomplete. Many hypotheses remain under study.

But humility does not mean surrendering reason.

The correct response to mystery is investigation, not worship.

Origin-of-life science shows that the ingredients and processes of life belong to nature. Organic molecules can form naturally. RNA can store information and perform chemical functions. Lipid-like molecules can form compartments. Energy gradients can drive reactions. Early Earth environments could provide heat, minerals, water, cycles, and chemical opportunity. Space itself can deliver some building blocks.

None of this proves every detail.

But it clearly shows that “God did it” is not required.

The origin of life is not evidence for divine creation. It is an open scientific question. And open scientific questions have a long history of becoming smaller as evidence grows.

The honest conclusion is simple:

We do not yet know every step by which life began.

But there is no scientific evidence that God was needed.

Chemistry came before biology.

Nature came before life.

And no divine spark has been demonstrated.

Chapter Summary

Evolution and the origin of life are related, but they are not the same question.

Evolution explains how life changes after life exists. Origin-of-life science studies how nonliving chemistry may have become biology.

The early Earth was a chemically active planet with water, minerals, energy sources, volcanic activity, impacts, and changing environments.

The building blocks of life are chemical, not supernatural.

Organic molecules can arise in natural settings on Earth and in space.

The RNA world hypothesis is one major scientific idea because RNA can store information and perform catalytic functions.

Protocells may have formed from simple membrane-like compartments.

Hydrothermal vents, hot springs, shallow ponds, wet-dry cycles, mineral surfaces, and molecules from space are all studied as possible parts of the origin-of-life story.

Science does not yet know every step from chemistry to the first life-like systems.

An unanswered scientific question is not evidence for God.

"God did it" does not provide a mechanism, make testable predictions, or guide research.

The origin of life remains a frontier of science, not proof of divine creation.

Key Takeaway

The origin of life is a profound scientific mystery, but mystery is not evidence of God. Science does not yet know every step by which chemistry became biology, but it has many natural pathways to investigate. Organic molecules, RNA, protocells, energy gradients, and early Earth environments all show that the question belongs to science. A gap in knowledge is not a miracle. No divine spark has been demonstrated.

Sources

NASA Astrobiology. Origin-of-Life and Prebiotic Chemistry Resources.

NASA. NASA's Asteroid Bennu Sample Reveals Mix of Life's Ingredients.

Bruce Alberts and colleagues. Molecular Biology of the Cell: The RNA World and the Origins of Life.

National Academies of Sciences, Engineering, and Medicine. Science, Evolution, and Creationism.

David Deamer and Bruce Damer. The Hot Spring Hypothesis for an Origin of Life.

J. Gómez-Márquez. The Origin of Life and Cellular Systems.

Addy Pross. What Is Life? How Chemistry Becomes Biology.

Nick Lane. The Vital Question.

John Maynard Smith and Eörs Szathmáry. The Major Transitions in Evolution.

Transition to the Next Chapter

Once life exists, the story enters a new stage.

Chemistry becomes heredity.

Heredity becomes variation.

Variation becomes selection.

Selection becomes evolution.

The next chapter will move from the origin of life to DNA, the molecule that records ancestry, carries inheritance, and reveals that living beings are not separate creations.

They are relatives.

Chapter 9: DNA: The Archive of Common Ancestry

The Family Secret Written in Every Cell

A child looks like his father.

The eyes may be similar. The smile may be similar. The shape of the face, the color of the skin, the texture of the hair, even certain medical risks may pass from one generation to the next.

Long before modern science, people noticed this. They did not know about DNA. They did not know about chromosomes. They did not know about genes. But they knew something powerful:

Life carries memory.

A child is not created from nothing. A child inherits. A body carries information from earlier bodies. A family has a biological history.

Modern genetics has taken that ordinary human observation and opened it into one of the strongest pieces of evidence for evolution.

DNA is not merely a molecule inside the cell. It is a record. It carries instructions, but it also carries ancestry. It contains working genes, regulatory regions, copied segments, repeated elements, damaged gene remnants, ancient viral traces, and inherited marks of the past.

If living beings were separately created by God, we might expect each kind of life to have its own independent design. We might expect clean biological separation. We might expect no deep pattern of shared genetic history.

But that is not what we find.

When we compare DNA across living beings, we find family relationships.

Humans are not genetically isolated from the rest of life. We share deep molecular connections with other animals, especially other primates. Mammals share genetic patterns with other mammals. Vertebrates share deeper patterns with other vertebrates. All known cellular life shares basic molecular machinery.

DNA does not point to separate divine creation.

DNA points to common ancestry.

From Resemblance to Evidence

Before DNA, people compared living beings by appearance.

They compared bones, teeth, organs, embryos, behavior, and fossils. They could see that a human arm, a bat wing, a whale flipper, and a cat leg share an underlying structure. They could see that mammals share certain features. They could see that humans resemble other primates in ways many people found uncomfortable.

But appearance alone has limits.

Some similarities arise because organisms live in similar environments, not because they are closely related. A shark and a dolphin both have streamlined bodies because both move through water, but one is a fish and the other is a mammal. A bird wing and a bat wing both allow flight, but their detailed anatomy reveals different histories.

DNA gives us a deeper tool.

It allows scientists to compare life at the molecular level. It allows us to ask whether the patterns seen in anatomy are also written in genes. It allows us to test common ancestry with evidence Darwin himself did not have.

The result is remarkable.

The tree suggested by anatomy, fossils, development, and geography is strongly supported by DNA.

This is why modern biology is so difficult for creationism. It is not only that fossils suggest evolution. It is not only that anatomy suggests evolution. It is not only that natural selection explains adaptation.

The genome itself carries the evidence.

What DNA Does

DNA is often described as the instruction molecule of life.

That is a useful beginning, but we should not treat DNA as magic. It is chemistry. It is made of atoms. It exists inside cells. It can be copied, damaged, repaired, rearranged, inherited, and changed.

DNA contains sequences that help guide the production of proteins and regulate biological processes. Genes are parts of DNA that contribute to inherited traits, often through the proteins they help produce or influence. Proteins then help build structures, run chemical reactions, carry signals, defend the body, move molecules, and perform countless tasks inside living systems.

In human beings, DNA is arranged in chromosomes. Children inherit DNA from their parents. That is why family resemblance is possible. That is also why some diseases run in families. Inherited changes in DNA can affect body structure, function, and risk of disease.

At the same time, DNA is not perfect.

It can mutate. It can be copied incorrectly. It can be damaged by radiation, chemicals, or viral infection. It can duplicate. Parts can be lost. Parts can be moved. Some changes matter greatly. Some do very little. Some are harmful. Some occasionally become useful under certain conditions.

This is exactly the kind of molecule evolution needs.

It preserves continuity, but it also allows change.

Without continuity, life could not inherit.

Without change, life could not evolve.

DNA Is Not a Heavenly Message

Religious apologists often say that DNA is a code, and every code requires a coder.

This sounds persuasive at first, but it is misleading.

DNA is sometimes called a code because sequences of DNA are interpreted by cellular machinery in ways that help build proteins. But scientific use of the word code does not automatically mean a supernatural mind wrote it.

We use similar language in other areas of nature. Tree rings contain information about age and growth conditions, but the tree was not keeping a diary. A snowflake has order, but no angel carved it. A crystal has structure, but no mind placed each atom by intention. Patterns in nature can carry information without being written by a conscious author.

DNA is a natural molecule. Its structure and function are studied through chemistry, biology, and genetics.

Calling DNA a divine code does not explain how genes work. It does not explain mutation. It does not explain inheritance. It does not explain why organisms share genetic errors. It does not explain why humans carry damaged gene remnants. It does not explain why our genome contains ancient viral sequences. It does not explain why DNA comparisons produce a branching tree of relationships.

The phrase "divine code" feels meaningful.

But it does no scientific work.

Evolution does.

The Genome as a Family Archive

Think of DNA as a family archive.

A family archive may contain birth records, old letters, photographs, copied documents, torn pages, spelling mistakes, repairs, and forgotten fragments from earlier generations. Some parts are useful. Some parts are damaged. Some parts no longer serve their original purpose, but they remain because they were inherited.

The genome is similar.

It contains functioning genes. It contains regulatory sequences. It contains duplicated genes. It contains repeated elements. It contains old viral remnants. It contains broken gene-like sequences. It contains sections that have been copied, changed, moved, silenced, or reused.

This is not what we would expect from a fresh, perfect blueprint created independently for each species.

It is exactly what we would expect from inheritance through deep time.

If a child and parent share a harmless spelling mistake in a copied family document, that shared mistake suggests inheritance. If cousins share the same old damaged document, that suggests they received it from a common ancestor. In the same way, shared genetic patterns, especially shared errors, are powerful evidence of common ancestry.

A perfect designer could make species similar.

But why make them share the same broken remnants in the same places?

Common ancestry explains this naturally.

They inherited them.

Shared Genes and Shared History

Many genes are shared across different forms of life.

This should not surprise us if life has common ancestry. Related organisms should share many genes because they inherited them from earlier ancestors. Closely related species should share more DNA than distantly related species. Distantly related species should still share deeper molecular features if they come from very ancient common ancestry.

That is what we find.

Humans share more genetic similarity with chimpanzees and bonobos than with dogs. We share more with dogs than with birds. We share more with birds than with insects. We share more with other animals than with plants. Yet all known cellular life shares basic features such as genetic material, protein-building machinery, and energy systems.

This is not a random pattern.

It is a nested pattern.

Nested patterns are exactly what we expect from family trees. Brothers and sisters resemble one another more than cousins. Cousins resemble one another more than distant relatives. Members of one family may share features that outsiders do not. Larger family groups share deeper ancestry.

The same logic applies to life.

DNA allows us to see that living beings are not separate divine products arranged side by side. They are related through ancestry.

The genome is not a wall separating life into created kinds.

It is a bridge connecting life across time.

Pseudogenes: Broken Genes With a Story

Pseudogenes are important genetic clues to evolution.

A pseudogene is a DNA sequence that resembles a gene but has lost some or all of its original function because of mutations or other changes. Some pseudogenes may have regulatory roles or other biological effects, so we should not casually call every pseudogene useless. Biology is often more subtle than old textbooks suggested.

But many pseudogenes are clearly damaged remnants of once-functional genes.

The important point is not simply that pseudogenes exist.

The important point is that related species often share similar pseudogenes in corresponding locations.

This is easier to explain by common ancestry than by separate creation.

Imagine two students submit essays with the same unusual spelling mistake in the same sentence. You would suspect they copied from a common source. If many students share the same strange error in the same place, common copying becomes the best explanation.

Shared pseudogenes work in a similar way.

If humans and other primates carry broken versions of the same gene in similar genomic locations, this strongly suggests inheritance from a common ancestor. Separate creation could say, "God put the same broken gene there." But that answer raises more questions than it solves.

Why would a perfect designer place shared broken remnants in related species?

Why would those remnants form the same branching pattern as anatomy and fossils?

Why would genetic mistakes fit evolutionary history so well?

Common ancestry explains the pattern simply.

The broken gene was inherited.

The Vitamin C Clue

One famous example involves vitamin C.

Many animals can make their own vitamin C. Humans cannot. We must obtain it from food. When humans lack enough vitamin C, they can develop scurvy, a disease once feared by sailors on long voyages.

The inability to make vitamin C is connected to a damaged gene involved in the vitamin C synthesis pathway. Humans are not alone in this. Other primates also carry a damaged version of the related gene.

This is exactly what common ancestry would predict.

If humans and other primates descended from ancestors in which this gene became damaged, their descendants would inherit the damage. If their diet provided enough vitamin C, the loss might not be fatal under ordinary conditions, so the broken gene could persist.

This is not elegant design.

A perfect designer could have allowed humans to make vitamin C, as many animals do. Instead, we carry a broken remnant and depend on diet. That makes sense through inheritance and evolutionary history.

It is harder to explain through perfect divine engineering.

Again, the issue is not whether a believer can invent a reason after the fact. The issue is which explanation actually fits the evidence better.

Evolution fits.

Endogenous Retroviruses: Ancient Infections in the Genome

Viruses can sometimes leave genetic traces in the cells they infect.

If viral DNA becomes inserted into reproductive cells, it can be passed to offspring. Over time, these viral remnants can become part of the inherited genome. These are called endogenous retroviruses.

This is one of the most striking pieces of genetic evidence for common ancestry.

If two species share the same viral insertion in the same genomic location, the simplest explanation is that the insertion occurred in a common ancestor and was inherited by both lineages.

The probability that similar viral sequences independently inserted into exactly the same genomic location in different species again and again is much less reasonable than common inheritance.

This is like finding the same old stain on copies of a family document.

It points to a shared source.

Some viral remnants may later be reused by evolution for new functions. Biology often modifies old material. That does not weaken the evidence for ancestry. It strengthens the larger point: genomes are historical, not freshly manufactured.

They carry scars, leftovers, and recycled pieces from the past.

A perfect divine blueprint would not need ancient viral fragments.

An evolved genome would be expected to carry them.

Human Chromosome 2: A Mark of Ancestry

Humans have 46 chromosomes. Chimpanzees, bonobos, gorillas, and orangutans have 48.

At first, this seems like a difference. But evolution makes a clear prediction. If humans and other great apes share common ancestry, and if our chromosome number differs, then something must have happened in the human lineage. One likely possibility is chromosome fusion.

That is what the evidence shows.

Human chromosome 2 contains features consistent with the fusion of two ancestral chromosomes. It has telomere-like sequences in an internal position, where chromosome ends appear to have joined. It also has signs of a second centromere region that is no longer active in the usual way.

This is powerful because it is not an argument from appearance. It is a genetic clue.

If humans were separately created, why would one of our chromosomes carry evidence of an ancestral fusion? Why would the pattern match the expectation from common ancestry with other apes?

Common ancestry explains it naturally.

Two ancestral chromosomes fused in the human lineage after it split from the lineage leading to chimpanzees and bonobos. The fused chromosome was inherited by descendants.

This does not make humans less valuable.

It makes our ancestry more visible.

The genome carries our history inside us.

Gene Duplication: New Possibilities From Old Material

Evolution often works by copying and modifying old material.

Gene duplication is one important way this happens. Sometimes a gene is copied. One copy may continue its original function, while the other copy is freer to accumulate changes. Over time, the duplicate may become nonfunctional, take on a modified function, or develop a new role.

This is not divine creation from nothing.

It is biological reuse.

Many important gene families appear to have histories involving duplication and modification. This helps explain how biological complexity can grow without needing a supernatural designer to invent every part separately.

A duplicated gene is like an extra copy of a tool. One copy can keep doing the old job. The other can be altered, damaged, or adapted for something different. Most changes may not lead to anything useful. But across deep time, some changes can matter.

This is how evolution often works.

It does not build with perfect originality.

It copies, modifies, repairs, breaks, silences, and reuses.

That is why genomes look historical. They contain families of related genes, old duplications, broken copies, and modified sequences.

A divine blueprint would not need this messy history.

Evolution predicts it.

DNA and the Tree of Life

When scientists compare DNA, they can build evolutionary trees.

These trees are not based on wishful thinking. They are based on patterns of similarity and difference. Closely related organisms tend to share more DNA. More distantly related organisms share less. Some genes are highly conserved because they perform basic functions. Others change more quickly.

The result is a branching pattern.

This pattern matches what evolution predicts. Species that share recent common ancestors cluster together. Species with more ancient common ancestors are farther apart. Genetic evidence can confirm, refine, or sometimes correct conclusions based on anatomy and fossils.

This is exactly how good science works.

It does not depend on one line of evidence. It tests ideas from multiple directions. If fossils, anatomy, embryology, geography, and DNA all point toward common ancestry, the conclusion becomes very strong.

Separate creation does not explain why independent sources of evidence form the same broad pattern.

It can only say, "God made it look that way."

But that is not science.

If all the evidence looks like common ancestry, behaves like common ancestry, and predicts like common ancestry, then common ancestry is the rational conclusion.

The Near-Universal Genetic Code

Nearly all known life uses the same basic genetic code, with some variations.

This means that the relationship between genetic sequences and amino acids is broadly shared across life. Cells also share basic systems for reading genetic information and building proteins.

This is a deep sign of life's unity.

If life were created in separate independent acts, there is no obvious reason every form should share such deep molecular machinery. A designer could have used entirely different systems for different kinds of organisms. Plants could have used one code, animals another, fungi another, bacteria another.

Instead, life shares a common molecular language.

Evolution explains this beautifully.

All known cellular life likely inherited core molecular machinery from ancient common ancestors. The shared code is not a divine signature. It is a family inheritance.

This does not mean every detail remained unchanged. Evolution modifies. Some organisms show variations. But the broad unity is powerful.

The deeper we look into life, the more connected it becomes.

DNA and Human Humility

The genetic evidence for common ancestry is especially humbling when it comes to humans.

Religious traditions often place humans in a separate category. We are said to be created in the image of God. We are said to possess a unique soul. We are said to stand above animals by divine decree.

DNA tells another story.

Human beings are mammals.

We are primates.

We are great apes.

We are related to other living beings through common ancestry.

This does not mean humans are identical to chimpanzees or any other animal. We are not. We have our own evolutionary history, language, symbolic thought, culture, science, medicine, music, literature, and moral imagination. But difference does not mean separate creation.

A person is different from his cousin, but still related.

Human uniqueness exists inside nature, not outside it.

This is a crucial point for science-based atheism. If humans are not separately created, then one of religion's central claims weakens. We no longer need God to explain our bodies, our ancestry, or our place in life.

We are not fallen angels.

We are evolved animals with extraordinary minds.

That truth is not degrading.

It is liberating.

Genetic Disease and the Imperfect Genome

DNA also reveals another serious problem for divine design:

The genome is not perfect.

Mutations can cause disease. Some mutations are inherited. Some occur during life. Some contribute to cancer. Some produce devastating childhood disorders. Some affect blood, nerves, muscles, metabolism, immunity, or development.

A perfect designer could create a flawless genome.

Evolution cannot.

Evolution works with copying, mutation, repair, selection, and inheritance. DNA replication is impressive, but it is not perfect. Repair systems exist, but they are not perfect. Harmful changes arise. Some are removed by selection. Some persist. Some appear again. Some remain hidden in carriers. Some cause disease only under certain conditions.

This is not what we would expect from perfect divine engineering.

It is exactly what we would expect from natural inheritance.

As a physician, one cannot ignore this. Genetic disease is not an abstract debate. It appears in families. It appears in children. It appears in cancers, inherited syndromes, metabolic disorders, blood diseases, and developmental problems.

If DNA is God's perfect code, why is it so vulnerable to harmful error?

If DNA is an evolved molecule, the answer is clearer.

It is powerful, but imperfect.

The Problem With Calling DNA Perfect

Some religious arguments praise DNA as perfect information.

But DNA is not perfect.

It contains mutations.

It contains broken gene remnants.

It contains repeated sequences.

It contains ancient viral insertions.

It contains duplicated segments.

It contains inherited vulnerabilities.

It can miscopy.

It can rearrange.

It can contribute to cancer and genetic disease.

This does not make DNA less fascinating.

It makes DNA more realistic.

DNA is not a polished divine manuscript. It is a living archive that has been copied for billions of years. It carries functional instructions, but it also carries errors, scars, and history.

This is one reason the design argument fails. It points to the impressive parts of DNA and ignores the messy parts. It praises the information and passes over the damage. It calls the useful gene a miracle and says little about the broken gene that causes suffering.

A fair explanation must include both.

Evolution does.

Perfect design does not.

Shared Errors Are Stronger Than Shared Successes

Shared useful features can sometimes be explained in more than one way.

A believer may argue that similar eyes, limbs, or organs were designed for similar purposes. That argument is weak, but it has emotional appeal.

Shared errors are harder for design to explain.

If two species share the same broken gene, the same viral insertion, the same unusual genetic marker, or the same chromosomal clue, common ancestry becomes far more convincing. It is one thing to say a designer reused a good design. It is another to say a perfect designer reused the same damaged pieces in the same places and arranged them in a branching pattern that matches evolution.

Shared mistakes reveal inheritance.

This is why DNA is so powerful.

It does not merely show that organisms are similar. It shows that they are similar in historically meaningful ways. They share not only functions, but scars.

A family can be recognized not only by healthy traits, but by inherited vulnerabilities, old records, shared documents, and repeated errors.

Life is the same.

The genome remembers.

Why Creationism Struggles With DNA

Creationism was easier to defend before genetics.

A person could look at a human, a chimpanzee, a dog, a whale, a bird, and a fish and say, "God made each kind separately." The similarities could be dismissed as divine design. The differences could be used to claim separation.

But DNA changed the debate.

Now we can compare life at the molecular level. We can examine sequences, chromosomes, genes, pseudogenes, viral remnants, mutations, and patterns of inheritance. We can build evolutionary trees. We can test relationships.

The evidence does not support separate creation.

It supports descent with modification.

Creationism can survive only by making God imitate evolution. God must make species look related. God must place shared genetic remnants in related species. God must arrange genetic patterns to match anatomy and fossils. God must create chromosome patterns that look like ancestral fusion. God must make genomes appear historical.

That kind of God is not a scientific explanation.

It is a way of refusing the obvious conclusion.

If the evidence looks like common ancestry, behaves like common ancestry, and predicts like common ancestry, then common ancestry is the rational conclusion.

DNA and the Failure of Special Creation

Special creation says that humans were made separately by God.

DNA says humans are related to other life.

Special creation says humans occupy a separate biological category.

DNA says humans are primates.

Special creation says our origin is supernatural.

DNA says our bodies carry the molecular marks of natural ancestry.

These two explanations do not have equal scientific strength.

One is based on scripture, tradition, and emotional identity.

The other is based on genetics, anatomy, fossils, development, and comparison.

This does not answer every philosophical or emotional question about human life. It does not tell us how to build justice, love, or meaning. But it does answer an important biological question:

Where do humans belong in life?

We belong inside the tree of life.

Not above it.

Not outside it.

Inside it.

This is a major step toward science-based atheism. If our bodies and genomes are products of natural ancestry, then we do not need divine creation to explain human life. God may remain as a private belief for some, but God is no longer a scientific necessity.

The evidence is written in every cell.

A More Honest Wonder

Some people fear that DNA reduces life to chemistry.

But life was always chemistry.

The question is not whether life is chemical. The question is whether chemistry can become organized, inherited, responsive, and conscious through natural processes.

Biology says yes.

DNA does not make life less wonderful. It makes life more astonishing. Inside each cell is a molecule carrying traces of ancient history. Inside the human body is evidence of kinship with other life. Every breath, every heartbeat, every thought, every act of love occurs in a body built from inherited biological information.

This is not spiritual emptiness.

It is natural grandeur.

The old religious story says God made humans from dust and breathed life into them.

Science tells a deeper story: atoms formed in stars, chemistry formed molecules, molecules formed life, life evolved cells, cells formed bodies, bodies formed brains, and brains became capable of asking where they came from.

That story does not need God.

It is already magnificent.

The Calm Conclusion

DNA has changed the argument about life.

Before genetics, the evidence for evolution came from fossils, anatomy, geography, development, and observed change. Those lines of evidence were already powerful. But DNA brought the evidence inside the cell.

The genome shows that life is inherited, not separately manufactured.

It shows that living beings carry shared molecular machinery. It shows that related species share more genetic similarity than distant species. It shows that pseudogenes, viral remnants, gene duplications, and chromosomal evidence fit common ancestry. It shows that humans are part of the primate family. It shows that the body's instructions are powerful, but imperfect.

This is not what we would expect from perfect divine design.

It is exactly what we would expect from evolution.

DNA is not a supernatural message.

It is a biological archive.

It does not say, "God made each creature separately."

It says, "Life is related."

No divine designer is required.

Chapter Summary

DNA carries inheritance and biological information, but it also carries evolutionary history.

Genes help shape biological traits, and mutations are natural changes in DNA sequence.

DNA comparisons allow scientists to test relationships among species.

Closely related organisms share more DNA than more distant organisms, forming a nested pattern of relationship.

Pseudogenes, shared genetic errors, endogenous retroviruses, and chromosomal clues strongly support common ancestry.

Human chromosome 2 contains evidence consistent with ancestral chromosome fusion.

Gene duplication shows how evolution can create new possibilities by copying and modifying old material.

The near-universal genetic code points to the deep unity of life.

Genetic disease and mutation show that DNA is not a perfect divine code.

DNA weakens special creation because it places humans inside the tree of life.

Key Takeaway

DNA is one of the strongest modern witnesses for evolution. It does not show separate divine creation. It shows inheritance, mutation, shared ancestry, broken gene remnants, ancient viral traces, and molecular kinship across life. The genome is not a perfect message from God. It is an archive of natural history written inside every cell.

Sources

National Human Genome Research Institute. Gene.

National Human Genome Research Institute. Mutation.

National Human Genome Research Institute. Genetic Code.

National Human Genome Research Institute. Comparative Genomics Fact Sheet.

National Human Genome Research Institute. Genetic Disorders.

National Academies of Sciences, Engineering, and Medicine. Science, Evolution, and Creationism.

National Academies Press. Evidence Supporting Biological Evolution.

University of California Museum of Paleontology. Understanding Evolution: Lines of Evidence.

J. W. IJdo and colleagues. Origin of Human Chromosome 2: An Ancestral Telomere-Telomere Fusion.

W. E. Johnson. Origins and Evolutionary Consequences of Ancient Endogenous Retroviruses.

Sean B. Carroll. The Making of the Fittest.

Jerry A. Coyne. Why Evolution Is True.

Richard Dawkins. The Greatest Show on Earth.

Neil Shubin. Your Inner Fish.

Transition to the Next Chapter

DNA shows that living beings are related.

But the next question is larger:

How are all these relationships arranged?

Are species separate divine products, each created in its own category? Or are they branches of one great living family?

The next chapter will examine the tree of life, the deep biological unity that connects bacteria, plants, animals, and human beings into one ancient story.

Chapter 10: The Tree of Life: One Family, Not Separate Creations

The Family Tree Bigger Than Humanity

Most families keep some kind of memory.

A photograph in a drawer. A name written in an old book. A story told by a grandparent. A face that appears again in a child born many years later. We understand family through resemblance, inheritance, and shared history.

Now imagine a family tree so large that it does not stop with parents, grandparents, and human ancestors.

It includes every human being. Then every ape. Then every mammal. Then every vertebrate. Then every animal. Then plants, fungi, bacteria, archaea, and the smallest forms of life. It reaches backward through oceans, forests, swamps, ice ages, volcanic landscapes, ancient seas, and a young Earth without humans, without religion, without language, and without prayer.

That is the tree of life.

It is one of the most powerful ideas in biology.

The tree of life says that living beings are not separate products placed into the world by divine command. They are related. They have history. They share ancestors. They are branches of one enormous biological family.

This idea changes everything.

It removes human beings from the center of creation and places us inside nature. It shows that life is not a showroom of divine designs, each made separately. It is a branching story of inheritance, change, survival, and extinction.

The living world does not look like separate creation.

It looks like ancestry.

The Old View: Separate Creations

For centuries, many religious traditions imagined life as a set of separate creations.

God made fish.

God made birds.

God made animals.

God made plants.

God made humans.

Each kind had its place. Each creature had its purpose. Human beings stood apart from all other life, often placed above animals by divine authority.

This picture was simple. It was emotionally satisfying. It made the world feel ordered and intentional. It gave human beings a special status. It made nature seem like a designed arrangement.

But simplicity is not the same as truth.

When biology began to examine life carefully, the old picture became harder to defend. Living beings did not fall into isolated boxes. They formed patterns. Some animals resembled one another deeply. Some shared bone structures. Some shared developmental pathways. Some shared genetic sequences. Some had fossil histories that connected them to earlier forms.

Life did not look like separate products.

It looked like relatives.

That difference is crucial.

Separate creation says similarity comes from the designer's choice.

Common ancestry says similarity comes from inheritance.

The evidence strongly supports inheritance.

The Pattern Within the Pattern

One of the most important facts in biology is that living beings form nested groups.

A human is an animal.

A human is also a vertebrate.

A human is also a mammal.

A human is also a primate.

A human is also a great ape.

A human is also Homo sapiens.

These groups are not random. They sit inside one another, like branches within larger branches.

This nested pattern is exactly what we expect from common ancestry.

It is also what we see in ordinary families. Siblings share recent parents. Cousins share grandparents. More distant relatives share older ancestors. The farther back we go, the larger the family becomes.

Biology works in the same way, but on a much larger scale.

Species that share a recent common ancestor tend to resemble one another more closely. Species with more distant common ancestors share deeper and older features. This produces a branching pattern.

Creationism struggles with this.

If God created each kind separately, why should life fall so consistently into nested relationships? Why should anatomy, fossils, DNA, and development all point toward the same branching history?

The answer is simple.

Life is not separately manufactured.

Life is related.

Homology: Similar Structures, Different Uses

One of the classic clues to common ancestry is homology.

Homology means that different organisms share structures because they inherited them from a common ancestor, even if those structures now serve different functions.

Consider the bones in a human arm, a bat wing, a whale flipper, and a cat front leg.

They do very different things.

The human arm can lift, hold, write, operate, embrace, and build.

The bat wing helps with flight.

The whale flipper helps with swimming.

The cat leg helps with walking and running.

Yet the underlying bone pattern is related. These are not entirely separate inventions. They are modified versions of an older limb structure inherited from common ancestors.

This is exactly what evolution predicts.

Evolution does not create every structure from nothing. It modifies old structures for new uses. The same basic limb plan can be changed across generations into a hand, wing, flipper, or leg.

Separate creation can only say, "God reused the same plan."

But that answer is weak. It does not explain why the pattern fits ancestry. It does not explain why fossil evidence also shows transitions. It does not explain why DNA supports the same relationships. It simply places a divine decision where evidence points to inheritance.

Homology is not the signature of a perfect designer.

It is the memory of common descent.

Convergent Evolution: Similar Solutions Without Close Kinship

Not every similarity means close ancestry.

This is important. A good scientific argument must be honest.

Sometimes unrelated or distantly related organisms evolve similar features because they face similar environmental pressures. This is called convergent evolution.

A shark and a dolphin both have streamlined bodies because both move through water. But a shark is a fish and a dolphin is a mammal. Their body shape is similar because water favors certain forms for swimming, not because they are close relatives.

Birds, bats, and insects all have wings, but their wings have different evolutionary histories. The wings perform similar functions, but they did not all come from the same recent winged ancestor.

Convergent evolution actually strengthens the case for evolution because it shows that natural selection can shape similar solutions under similar pressures.

It also shows the difference between surface appearance and deeper history.

A casual observer may see a shark and dolphin and think they belong together. Biology looks deeper. It examines bones, reproduction, development, DNA, and ancestry. A dolphin breathes air, nurses its young, has mammalian anatomy, and shares ancestry with other mammals. A shark has a different history.

This is why science is more reliable than appearance.

The tree of life is not built from superficial resemblance alone. It is built from many lines of evidence.

The Unity Beneath Diversity

Life is astonishingly diverse.

A mushroom does not look like an eagle. A bacterium does not look like a rose. A whale does not look like an oak tree. A human being does not look like a yeast cell.

But beneath this diversity, there is deep unity.

Living cells use chemical energy. They contain genetic information. They build proteins. They use complex molecular machinery. They reproduce. They interact with their environments. They are made of the same basic chemistry.

This unity is one of the strongest signs that life has a shared history.

If life were created in separate divine acts, there is no obvious reason for such deep molecular unity. A designer could have made plants with one kind of information system, animals with another, fungi with another, and bacteria with another. A divine engineer could have used completely different biological languages.

But life does not look like that.

Life shares a common molecular foundation.

This makes sense if all known life descends from ancient ancestors. The details changed over time, but the deepest systems were inherited.

The unity of life is not proof of a divine artist.

It is evidence of a biological family.

LUCA: The Last Universal Common Ancestor

Scientists use the term LUCA for the Last Universal Common Ancestor.

LUCA does not mean the first life.

This is important.

LUCA means the last common ancestor of all living organisms known today. There may have been earlier life forms before LUCA. There may have been many branches that died out. LUCA is the point from which all current cellular life appears to descend.

This idea is profound.

It means that bacteria, archaea, plants, animals, fungi, and human beings all share a deep ancestral connection. The connection is very ancient, but it is still written in the shared features of life.

All known cellular life uses genetic information. All known cellular life uses molecular machinery to build proteins. All known cellular life depends on energy flow and chemical organization. These shared features suggest deep inheritance.

LUCA was not human-like. It was not an animal. It was not a plant. It was a very ancient cellular or cell-like ancestor with molecular systems that later life inherited and modified.

The idea of LUCA replaces separate creation with shared biological origin.

It says that life is not many unrelated miracles.

It is one ancient branching story.

The Tree Is Not Always Simple

The tree of life is a powerful image, but nature is more complex than a simple drawing.

Especially among microbes, genes can sometimes move sideways between organisms. This is called horizontal gene transfer. Bacteria, for example, can exchange genetic material in ways that do not follow the simple parent-to-child pattern seen in animals.

This means that the deepest history of life is sometimes more like a web than a neat tree.

But this does not help creationism.

It does not show separate divine creation. It shows that life's natural history is even more complex, dynamic, and interconnected than a simple tree can fully capture.

For plants and animals, the tree pattern remains very useful. For microbes, the pattern includes both branching ancestry and gene exchange. Biology is honest enough to refine its images when evidence requires it.

That is the strength of science.

Science does not worship its own diagrams. It improves them.

The tree of life remains a powerful symbol, but we should understand it maturely. Life branches, and some branches also exchange genetic material. The broad conclusion remains the same:

Life has natural relationships.

It does not appear as separate divine products.

Classification Makes Sense Because of Ancestry

Why does biological classification work?

Why can we place humans with apes, apes with primates, primates with mammals, mammals with vertebrates, and vertebrates with animals?

In the old creation view, classification was often seen as discovering the order in God's mind. God made categories, and human beings named them.

Evolution gives a better explanation.

Classification works because life has ancestry. Organisms fall into groups because they are related. Mammals share features because they inherited them from mammalian ancestors. Primates share features because they inherited them from primate ancestors. Humans and other great apes share features because they are close relatives.

The deeper the ancestry, the broader the group.

This is why modern classification is not just naming. It is history.

A species is not merely placed in a category. It is located on a branch of life's tree.

This changed biology completely. It turned classification from a filing system into a record of descent.

Separate creation can arrange organisms into groups, but it cannot explain why those groups have the deep structure of ancestry.

Evolution can.

Humans Inside the Tree

The tree of life is emotionally difficult for many religious people because it places humans inside nature.

Not above it.

Not outside it.

Inside it.

Human beings are animals. More specifically, we are mammals, primates, and great apes. This does not mean we are the same as every other animal. It does not mean humans have no special qualities. We have language, symbolic thought, moral imagination, science, medicine, art, music, and complex culture.

But our uniqueness is natural, not supernatural.

A bird is unique in its own way. A whale is unique in its own way. An octopus is unique in its own way. A human being is unique in a human way. Uniqueness does not require separate creation.

Religion often teaches that humans are special because God made us in His image. Biology teaches that humans are special because evolution produced a social, intelligent, language-using primate with a large brain and extraordinary cultural abilities.

This second explanation is humbler, but it is more honest.

We do not need to be magically separate to matter.

We matter because we can suffer, love, think, remember, care, and understand. We matter because life is fragile. We matter because consciousness is precious. We matter because we matter to one another.

The tree of life does not degrade humanity.

It connects humanity.

The Problem for Special Creation

Special creation says that humans were created separately.

The tree of life says otherwise.

If humans were separately created, why do we fit so clearly within the primate branch? Why do we share anatomy with other apes? Why do we share genetic markers? Why do we share embryological features? Why do we have the same basic mammalian body plan? Why does the fossil record show a branching history of human relatives?

Creationism can always answer, "God chose to make humans that way."

But that is not a scientific explanation. It explains every possible outcome and therefore explains nothing.

If humans looked completely different from all other life, creationists would call that evidence of special creation.

But humans look biologically related to other animals, and creationists call that design too.

If both evidence and its opposite are called proof of God, then the argument is not scientific.

Evolution, by contrast, makes clear expectations. It predicts that humans should fit into the tree of life. It predicts that our anatomy, DNA, development, and fossils should show connection to other primates and mammals.

They do.

The evidence points toward natural ancestry, not special creation.

One Tree, Many Dead Branches

The tree of life is not only a tree of living species.

It is also a tree of extinction.

Many branches have ended. Some ended quietly. Some ended in mass extinction. Some left descendants. Many left none. The life we see today is only a small surviving part of a much larger history.

This matters because it weakens the idea of divine planning.

A religious picture of creation often imagines life as ordered, purposeful, and arranged. But the actual history of life is filled with failed branches, vanished species, and lost worlds.

The tree of life is not a carefully pruned garden designed for human arrival.

It is a wild history of growth, struggle, accident, survival, and death.

Some branches survived because they were well adapted. Some survived because they were lucky. Some vanished despite being complex and impressive. Some simple forms survived for billions of years.

There is no moral ladder.

There is no clear divine staircase.

There is branching life and extinction.

This is not what most people imagine when they say nature proves God.

Common Ancestry Explains Both Beauty and Imperfection

The tree of life helps explain why living beings can be beautiful and imperfect at the same time.

Beauty comes from adaptation, diversity, and life's long creative history. Imperfection comes from the fact that evolution modifies what already exists. It does not design from nothing.

A bat wing, whale flipper, and human arm are beautiful in their different ways. But they are also constrained by history. They are modifications of older structures, not perfect inventions created independently for each task.

The same is true throughout life.

The body carries history.

The genome carries history.

Development carries history.

Behavior carries history.

This is why life does not look like perfect engineering when examined closely. It looks like ancestry.

A designer could create the best possible structure for each purpose. Evolution reworks inherited structures and keeps what functions well enough.

Common ancestry therefore explains both the unity and the messiness of life.

It explains why living beings are related.

It also explains why they are not perfect.

The Spiritual Temptation of Human Centrality

One reason people resist the tree of life is that it removes human beings from the center.

Religious stories often place humans at the heart of creation. The world exists for us. Animals exist beneath us. Nature exists as our stage. God's plan revolves around human destiny.

The tree of life challenges that deeply.

It shows that humans arrived late. It shows that life existed for billions of years without us. It shows that many branches flourished and vanished before human beings appeared. It shows that our bodies are related to other animals. It shows that we are one branch, not the trunk of the whole tree.

This can feel like a loss.

But it can also be a moral awakening.

If we are part of nature, not rulers placed above it by divine privilege, then our relationship to other life must become more responsible. We are not outside the web of life. We depend on it. We emerged from it. We share ancestry with it. We damage it at our own risk.

Science-based atheism does not make humans less responsible.

It makes us more responsible.

There is no divine parent coming to repair the damage.

The care of life belongs to us.

Why the Tree of Life Needs No God

Some believers may accept the tree of life but say God created the tree.

That is a theological belief, not a scientific conclusion.

Science can study the tree of life without adding God. It can study fossils, DNA, anatomy, development, speciation, extinction, and common ancestry. It can explain the branching pattern through natural processes.

Adding God does not improve the explanation.

It does not explain why life shares molecular machinery.

It does not explain why species fall into nested groups.

It does not explain why fossils appear in historical sequence.

It does not explain why bodies show inherited structures.

It does not explain why genomes contain shared errors.

It does not explain extinction.

If God is placed behind the whole process but leaves no testable mark, makes no measurable difference, and explains nothing better than evolution, then God is not a scientific explanation.

This is not an insult.

It is a boundary.

Science follows evidence. It does not add invisible causes where natural causes are sufficient.

The tree of life is explained by common ancestry and evolution.

No divine gardener is required.

The Tree as a Better Story

The old creation story gives a simple answer:

God made each kind.

The tree of life gives a deeper answer:

Life is related through descent, change, and time.

This scientific story is not cold. It is deeply moving.

It means the bird, the tree, the whale, the dog, the mushroom, the bacterium, and the human child all belong to one long history. It means life is not a set of strangers. It is one extended family. It means the living world is not a collection of objects placed around us. It is our own larger kinship.

The tree of life also gives us humility.

We are not the owners of life. We are participants in it.

We are not the reason for all branches. We are one branch.

We are not separate from nature. We are nature with memory, language, science, and moral responsibility.

That story does not need God to be meaningful.

It is meaningful because it is true.

The Calm Conclusion

The tree of life is one of the clearest answers to separate creation.

Life does not appear as isolated divine products. It appears as a branching system of relationships.

Anatomy shows it.

Fossils show it.

DNA shows it.

Development shows it.

Classification shows it.

The shared machinery of cells shows it.

Species are not fixed categories placed separately into the world. They are historical populations with ancestors and relatives.

Humans are not outside this tree. We are part of it. We belong to the same living history as every other organism on Earth.

This truth may challenge religious pride, but it does not destroy human dignity. It gives us a more honest dignity. We are evolved life capable of understanding evolution. We are nature becoming conscious of its own history.

The tree of life does not show scientific evidence of God.

It shows common ancestry.

It shows natural history.

It shows that life is one family, not separate creations.

No designer is required.

Chapter Summary

The tree of life shows that living beings are related through common ancestry.

The old view of separate creation imagined species as independent divine products, but biology reveals nested relationships.

Homology shows that different organisms can use modified versions of inherited structures for different purposes.

Convergent evolution shows that similar environments can shape similar traits in unrelated or distant lineages.

The deep molecular unity of life supports shared ancestry.

LUCA, the Last Universal Common Ancestor, represents the ancient ancestor from which all known cellular life descends.

The tree of life can be more web-like among microbes because of horizontal gene transfer, but this still reflects natural history, not separate creation.

Humans belong inside the tree of life as mammals, primates, and great apes.

The tree of life weakens special creation and supports science-based atheism by showing that life is connected through natural ancestry.

Key Takeaway

Living beings are not separate divine products. They are branches of one ancient tree of life. Anatomy, DNA, fossils, development, and cellular biology all point toward common ancestry. Humans are not outside this tree. We are part of it. The living world does not reveal separate creation by God. It reveals one vast family shaped by evolution, inheritance, and time.

Sources

Charles Darwin. On the Origin of Species.

National Academies of Sciences, Engineering, and Medicine. Science, Evolution, and Creationism.

University of California Museum of Paleontology. Understanding Evolution: The Tree of Life Today.

University of California Museum of Paleontology. Understanding Evolution: Homologies and Analogies.

National Human Genome Research Institute. Comparative Genomics Fact Sheet.

G. P. Fournier and colleagues. Ancient Horizontal Gene Transfer and the Last Common Ancestor.

V. Daubin and G. Szollosi. Horizontal Gene Transfer and the History of Life.

M. C. Weiss and colleagues. The Last Universal Common Ancestor Between Ancient Earth Chemistry and the Onset of Genetics.

Neil Shubin. Your Inner Fish.

Jerry A. Coyne. Why Evolution Is True.

Richard Dawkins. The Greatest Show on Earth.

Transition to the Next Chapter

The tree of life shows that living beings are related.

But relationship is not the only evidence against divine design. When we look closely at bodies themselves, we find something equally powerful.

The body is not a perfect machine.

It is a collection of inherited structures, compromises, detours, and old parts used in new ways.

The next chapter will examine why life looks less like engineering and more like tinkering.

Chapter 11: Tinkering, Not Engineering

The Back That Bent Too Far

A man bends down to lift a small box from the floor.

The box is not heavy. He has lifted heavier things many times. But this time, something catches in his lower back. A sharp pain spreads across his spine. He freezes, half bent, afraid to move. For the next several days, every step, cough, turn, and attempt to stand reminds him of a simple truth:

The human body is remarkable, but it is not perfectly designed.

Anyone who has seen enough illness, injury, birth, aging, and pain knows this.

The body heals wounds. It fights infection. It balances salt, water, acid, temperature, and blood pressure. It turns food into energy. It allows movement, speech, memory, thought, love, work, and reproduction. It is a living system of astonishing complexity.

But it is also fragile.

The spine supports upright walking, but it often fails under ordinary strain. The throat allows speech and swallowing, but its structure makes choking possible. The eye sees beautifully, but it has a blind spot. The pelvis allows upright posture, but it makes childbirth difficult and sometimes dangerous. Teeth help us chew, but wisdom teeth often crowd the jaw. The appendix may have immune-related functions, but it can become inflamed and rupture. The immune system protects us, but it can also attack our own body.

This is not the pattern of flawless engineering.

It is the pattern of evolution.

Life was not designed from a clean drawing board. It was modified from older life. The body is not a perfect machine created from nothing. It is a historical structure, shaped by inheritance, change, compromise, and survival.

Evolution does not engineer from the beginning.

It tinkers with what already exists.

The Engineer and the Tinkerer

A good engineer can begin with a goal and then design the best possible structure for that goal.

If the task is to build a bridge, the engineer can calculate weight, materials, stress, safety, and future use. If the task is to build an airplane, the engineer can study airflow, fuel efficiency, strength, and maintenance. If the design is unsafe, the engineer can go back to the drawing board.

Evolution has no drawing board.

It cannot begin again from zero. It cannot plan for a distant future. It cannot say, “This spine will cause trouble later, so let us redesign the whole skeleton.” It cannot say, “This birth canal is dangerous, so let us rebuild the pelvis.” It cannot say, “This nerve route is inefficient, so let us run it directly.”

Evolution works only with what already exists.

It modifies.

It reuses.

It adjusts.

It patches.

It compromises.

It preserves what works well enough.

This is why living bodies often resemble old houses that have been repaired many times. A wall was moved. A pipe was extended. A wire was rerouted. A room was added. A stairway became awkward. The house still works, but it carries its history inside its structure.

The body is similar.

It works in the present, but it carries the past.

Useful Does Not Mean Perfect

One of the great mistakes in the design argument is confusing usefulness with perfection.

The eye is useful, therefore it must be perfectly designed.

The hand is useful, therefore it must be divinely planned.

The heart is useful, therefore it must be the work of God.

The spine is useful, therefore it must be ideal.

But useful does not mean perfect.

A structure can work well enough and still be flawed. A car with poor fuel efficiency can still drive. A house with awkward plumbing can still shelter people. A road with sharp turns can still carry traffic. A body with weak points can still survive and reproduce.

Natural selection does not need to create the best possible body.

It only needs to preserve traits that help survival and reproduction enough in a particular environment.

This explains why so many biological structures are impressive but imperfect. They did not arise because an all-knowing designer chose the best possible arrangement. They arose because earlier structures were modified over time, and the results were good enough to continue.

Good enough is not divine perfection.

Good enough is evolution.

The Human Spine: Upright Walking With a Price

The human spine is one of the clearest examples of evolutionary compromise.

Our ancestors did not begin as upright walkers. The vertebrate body has a long history. Human beings inherited a spine shaped by earlier forms of life, and that spine was later modified for bipedal walking. Standing upright brought advantages. It freed the hands. It allowed carrying, tool use, long-distance movement, and a wider view of the surroundings.

But the change came with costs.

The human spine must support the weight of the upper body while balancing on two legs. It curves. It compresses. It bears stress. Discs can degenerate or herniate. Nerves can be pinched. Muscles can spasm. Joints can wear down. Lower back pain is one of the most common human complaints.

A perfect engineer designing an upright intelligent creature from scratch might have chosen a more stable support system.

Evolution did not start from scratch.

It modified an inherited vertebrate spine.

This does not mean the spine is useless. It is highly useful. It allows flexibility, posture, protection of the spinal cord, and movement. But its weaknesses are exactly what we should expect from evolutionary modification.

It is not a pillar designed for perfect upright life.

It is an old structure carrying a new burden.

The Throat: Speech With a Risk of Choking

Human speech is one of our most remarkable abilities.

With air, vocal cords, tongue, lips, throat, and brain, we produce language. We tell stories. We teach children. We express love, warning, memory, humor, grief, and science. Speech is central to human culture.

But the same anatomy that helps speech also creates danger.

In humans, the passages for air and food cross in a way that creates the risk of choking. Food must pass toward the esophagus, while air must pass toward the trachea. A small flap, the epiglottis, helps protect the airway during swallowing. Most of the time this system works.

But not always.

A piece of food can enter the airway. Choking can kill.

This is a strange arrangement for perfect design.

Why would a wise designer place eating and breathing in such a shared and risky passage? Why not design fully separate channels from the beginning? Why build a system where one wrong movement during an ordinary meal can block oxygen to the brain?

Evolution explains this better.

The human throat is a modified structure with a history. Speech, swallowing, breathing, posture, head shape, and laryngeal position all interact. The system works well enough most of the time, but it is not ideal.

Again, useful does not mean perfectly designed.

The Eye: Beautiful Vision With a Blind Spot

The human eye is often used as evidence for God.

And it is easy to understand why.

The eye receives light, focuses images, adjusts to brightness, detects color, and gives us a visual world rich in detail. A person can see a face, read a book, admire the ocean, watch a bird in flight, and recognize danger. Vision is one of life's great gifts.

But the vertebrate eye has an important flaw.

The retina is arranged so that nerve fibers pass in front of the light-sensitive cells and then gather to form the optic nerve. Where the optic nerve exits the eye, there are no light-sensitive cells. This creates a blind spot.

The brain fills in the missing area so smoothly that we usually do not notice it. But the blind spot is real.

This is not the cleanest possible arrangement.

In cephalopods such as octopuses, the eye is arranged differently. Their photoreceptors face the incoming light in a non-inverted retina, and they do not have the same kind of blind spot caused by the optic nerve passing through the retina.

This comparison matters because it shows that biological vision can be built in more than one way.

A perfect designer could have avoided the blind spot.

Evolution explains why it exists.

The vertebrate eye developed through historical pathways. Once a structure exists, evolution modifies it. It does not easily reverse everything and rebuild from an ideal plan.

The eye is wonderful.

But it is not perfect engineering.

It is evolved vision.

The Recurrent Laryngeal Nerve: A Long Detour

The recurrent laryngeal nerve is often mentioned as an example of evolutionary tinkering.

In mammals, this nerve travels from the brain down into the chest, loops around major blood vessels, and then returns upward to the larynx. In humans, this route is longer than a direct path would be. In giraffes, the detour is far more dramatic because the neck is so long.

A direct engineer would likely route the nerve more simply.

So why does it take such a strange path?

Evolutionary history gives the answer.

The nerve's route makes sense when understood through development and descent from earlier vertebrate anatomy. As bodies changed, necks lengthened, and the heart and major vessels shifted during development, the nerve remained constrained by its anatomical relationships. Evolution stretched the route rather than redesigning it from scratch.

This is exactly the kind of thing we expect from tinkering.

A tinkerer works around old arrangements. An engineer starting fresh could choose a cleaner path. Evolution cannot simply erase the old developmental plan and begin again.

The recurrent laryngeal nerve does not disprove every possible God by itself. No single anatomical example can do that.

But it strongly challenges the claim that bodies show perfect divine engineering.

Its route looks like history.

Not ideal design.

Wisdom Teeth and the Shrinking Jaw

Wisdom teeth are another ordinary example of evolutionary mismatch.

Many people do not have enough space in the jaw for their third molars. Wisdom teeth may become impacted, painful, infected, or damaging to nearby teeth. Dentists and oral surgeons often remove them.

Why would a perfect designer create teeth that commonly do not fit?

Evolution gives a better explanation.

Our ancestors had different diets, different jaw use, and often larger jaws. Over time, changes in diet, cooking, food processing, and jaw development affected the space available for teeth. The third molars remained, but the jaw often became too small to accommodate them comfortably.

This is not a perfectly coordinated design.

It is a historical mismatch.

The body carries structures that made more sense in earlier conditions but may cause problems now. Evolution does not remove every troublesome feature quickly, especially if the harm often occurs after reproductive age or if the feature is not harmful enough to strongly affect survival and reproduction.

Wisdom teeth are not a deep mystery.

They are old parts in a changed body.

The Appendix: Not Useless, But Vulnerable

The appendix is often described as useless.

That is too simple.

The appendix may have immune-related roles and may serve as a reservoir for helpful gut bacteria. Biology is rarely as simple as "useful" or "useless."

But the appendix still creates a serious design problem.

It can become inflamed. Appendicitis can cause severe pain, infection, rupture, peritonitis, sepsis, and death if untreated. Before modern surgery and antibiotics, appendicitis was far more dangerous.

A structure can have some function and still be poorly designed from the standpoint of safety.

This is important because defenders of design often respond to every example of imperfection by finding some possible function. But the existence of some function does not prove perfect design.

A dangerous bridge may still carry traffic.

That does not make it well engineered.

The appendix may have biological roles, but its vulnerability to life-threatening inflammation fits better with evolutionary history than perfect divine planning.

The body is not built to avoid every danger.

It is built by inherited compromise.

Childbirth: A Difficult Evolutionary Bargain

Human childbirth is beautiful, but it can also be dangerous.

The human baby has a large head because of brain development. The human pelvis is shaped by the demands of upright walking. These two facts create a difficult passage. Labor can be prolonged. The baby can become obstructed. The mother can hemorrhage. Before modern obstetrics, maternal and infant death were tragically common.

Why would a loving designer make childbirth so painful and dangerous?

Religion often praises childbirth as a miracle, but it rarely faces the full biological cost. A newborn child can inspire awe, but an honest view must also remember obstructed labor, fistulas, hemorrhage, sepsis, stillbirth, birth injury, and maternal death.

Evolution explains the problem through trade-offs.

Upright walking changed the pelvis. Brain expansion increased head size. Natural selection did not create a perfect solution. It preserved a workable compromise, later assisted by social care, culture, midwifery, surgery, antibiotics, blood transfusion, and modern medicine.

This is not ideal engineering.

It is a difficult evolutionary bargain.

The existence of human life does not prove divine kindness. The danger built into reproduction raises serious questions for perfect design.

The Knee, the Foot, and the Price of Walking

Human beings walk upright with remarkable efficiency. Our feet, legs, hips, and knees allow us to move long distances, stand, run, carry, and work with our hands free.

But these structures also fail.

Knees wear down. Ligaments tear. Feet develop pain, bunions, flat arches, plantar fasciitis, and other problems. Hips fracture in old age. Balance becomes fragile. Joints carry years of mechanical stress.

Again, the issue is not that the body does nothing well.

It does many things well.

The issue is that it does not look like flawless design.

It looks like an inherited structure modified for upright walking, carrying both advantages and costs. Bipedalism gave our ancestors major benefits, but it did not produce perfect durability.

Evolution works with trade-offs.

Every gain can bring a cost.

That is not divine engineering.

It is natural history.

Goosebumps and Other Leftovers

Goosebumps are a small but revealing example.

In many mammals, tiny muscles at the base of hairs can raise the hair, making the animal appear larger or helping trap air for warmth. In humans, with much less body hair, the response remains but is far less useful. We still get goosebumps when cold or emotionally stirred, but the effect does little compared with what it does in fur-covered animals.

This is a minor example, but it tells a larger story.

The human body contains traces of ancestry. Some features have reduced function. Some have changed function. Some are leftovers. Some are modified. Some remain because they do not cause enough harm to be removed by selection.

Creation thinking often imagines bodies as fresh designs.

Evolution sees them as historical documents.

Goosebumps do not prove atheism by themselves. No single example does. But together with many other features, they show a pattern.

We are not separate creations.

We are modified descendants.

The Mammalian Ear: Old Bones in a New Role

The tiny bones in the mammalian middle ear are beautiful examples of evolutionary transformation.

Mammals have three small ear bones: the malleus, incus, and stapes. These bones help transmit sound. Fossil and developmental evidence show that some of these bones are related to bones that formed part of the jaw structure in earlier vertebrates.

This is a profound idea.

A structure used in one role can be modified over evolutionary time into a structure serving a different role.

This is not what one would expect from an engineer designing every system independently from scratch. But it is exactly what one expects from evolution. Old parts can be repurposed. A bone once related to jaw function can become part of hearing.

Evolution is full of such reuse.

Bird wings are modified forelimbs. Whale flippers are modified mammalian limbs. Human hands are modified vertebrate forelimbs. The middle ear contains transformed elements with deep ancestry.

Life is not a set of separate inventions.

It is a history of modification.

The Body as a Living Museum

The human body is not only a functioning organism.

It is also a living museum.

The spine carries the history of vertebrates and upright walking.

The hand carries the history of limbs.

The ear carries the history of older jaw structures.

The eye carries the history of vertebrate development.

The pelvis carries the tension between walking and childbirth.

The skin, muscles, nerves, and organs carry ancient biological patterns.

The genome carries the history of ancestry, duplication, mutation, and inheritance.

A museum contains objects from different times. Some are beautiful. Some are damaged. Some are reused. Some are preserved because they tell a story.

The body is like that.

It works in the present, but it is built from the past.

This is why evolution explains the body so well. It does not expect perfect design. It expects modified inheritance. It expects compromise. It expects traces of earlier forms. It expects vulnerability.

Perfect design must explain why the body looks so historical.

Evolution simply says:

Because it is.

Why Bad Design Arguments Must Be Used Carefully

We must be careful.

The argument is not that every imperfect structure proves there is no God. That would be too simple. A religious believer can always say that God had unknown reasons or that human beings cannot understand divine wisdom.

The stronger argument is different.

The living body looks exactly as we would expect if evolution were true.

It does not look the way we would reasonably expect if a perfect, all-knowing, all-powerful, loving designer had built life from nothing.

This is a more precise argument.

It does not depend on one example. It depends on a pattern.

Spine problems, choking risk, blind spots, nerve detours, wisdom teeth, appendicitis, dangerous childbirth, joint failure, genetic disease, and anatomical leftovers all point in the same direction. The body is functional, but not ideal. It is impressive, but not perfect. It is adapted, but historically constrained.

A perfect designer could have done better.

Evolution could only work with what it inherited.

That is the point.

Why "God Wanted It That Way" Is Not Science

Some people respond to every flaw by saying, "God wanted it that way."

But this explains nothing.

Why does the eye have a blind spot?

God wanted it.

Why does childbirth involve danger?

God wanted it.

Why does the nerve take a long detour?

God wanted it.

Why do children get genetic disease?

God wanted it.

Why do bodies age and fail?

God wanted it.

This kind of answer may protect faith, but it does not help understanding. It does not explain anatomy. It does not explain development. It does not explain fossils. It does not explain genetics. It does not explain disease. It does not make testable predictions.

It only stops the question.

Science does not stop.

Science asks how structures developed, what ancestry they reflect, what constraints shaped them, what trade-offs they carry, and why they persist.

That approach produces knowledge.

It leads to medicine, surgery, prevention, genetics, physical therapy, obstetrics, dentistry, and a deeper understanding of the body.

The divine design answer produces no comparable scientific progress.

The Medical View of the Body

The body looks different when seen through illness.

From the outside, it is easy to praise the healthy body as divine design. But medicine forces us to see the whole body, not only its successes.

There are ruptured appendixes, obstructed airways, herniated discs, autoimmune diseases, cancers, congenital anomalies, kidney failure, heart attacks, strokes, sepsis, fractures, and degenerative joints.

This does not make the body less impressive.

It makes the design argument less convincing.

The body is a wonderful evolved system, but a poor example of perfect engineering. It is powerful, but vulnerable. It heals, but not always. It protects, but sometimes harms itself. It reproduces, but with risk. It adapts, but with trade-offs. It survives, but only temporarily.

Disease and anatomy are not separate from the argument about God. They are central to it.

If someone says the body proves a loving designer, the honest question is:

Which body?

The healthy body only?

The child with a fatal genetic disorder?

The mother dying in childbirth?

The patient choking on food?

The elderly person crippled by spinal disease?

The young person with leukemia?

The infant born with a malformed heart?

A fair argument must include all of them.

Evolution can.

Perfect design struggles.

Tinkering Explains More

The idea of evolutionary tinkering explains the body better than perfect design.

It explains why old structures are reused.

It explains why organs are good enough but not ideal.

It explains why different species share body plans.

It explains why anatomy carries traces of ancestry.

It explains why some features persist despite causing problems.

It explains why trade-offs are everywhere.

It explains why bodies are vulnerable to disease and failure.

Tinkering is not a poetic insult. It is a scientific description of how evolution works. It uses available variation. It modifies earlier structures. It cannot see the future. It cannot rebuild from scratch. It cannot create perfection. It preserves what works well enough under existing conditions.

This is why life is both impressive and flawed.

The design argument sees the impressive parts and calls them God.

Evolution sees the whole pattern and explains both success and failure.

That is why evolution is the stronger explanation.

The Body Without Worship

We do not need to worship the body as divine design in order to respect it.

In fact, understanding the body as evolved may deepen our respect.

The body is not a magical object. It is the result of billions of years of life, survival, adaptation, and inheritance. It connects us to fish, reptiles, mammals, primates, and ancient cells. It carries the memory of old environments and old struggles. It is fragile because it is natural. It is precious because it is temporary.

This view does not cheapen human life.

It makes human life more tender.

If there is no perfect designer watching over the body, then medicine matters more. Prevention matters more. Compassion matters more. Science matters more. Surgery, vaccines, antibiotics, sanitation, genetics, nutrition, public health, and careful care matter more.

We are not protected by a divine engineer.

We must protect one another.

That is a mature, humane conclusion.

The Calm Conclusion

The human body is remarkable, but it is not perfect.

It is not a flawless machine made from a clean divine blueprint. It is a living product of evolutionary history. It contains old structures used in new ways, useful systems with dangerous weaknesses, and compromises shaped by survival rather than ideal design.

The spine supports us, but it fails.

The throat gives us speech, but it can choke us.

The eye sees, but it has a blind spot.

The nerve travels an unnecessary detour.

The jaw often lacks room for wisdom teeth.

The appendix can threaten life.

Childbirth can be dangerous.

The immune system can attack the body.

Cells that divide can become cancer.

This pattern is not what we would expect from perfect divine engineering.

It is exactly what we would expect from evolution.

The body is not evidence of God.

It is evidence of ancestry, adaptation, compromise, and history.

Life was not perfectly engineered.

It was tinkered into being.

No divine designer is required.

Chapter Summary

The human body is remarkable, but it is not perfectly designed.

Evolution works by modifying existing structures, not by designing from a clean blueprint.

Useful does not mean perfect. Many body parts work well enough but still carry serious flaws.

The human spine reflects the compromise of upright walking and commonly causes pain.

The throat allows speech, but its shared pathway for food and air creates a risk of choking.

The vertebrate eye sees well, but it has a blind spot.

The recurrent laryngeal nerve follows a long detour that makes sense through developmental and evolutionary history.

Wisdom teeth, appendicitis, difficult childbirth, joint problems, and goosebumps all show signs of historical compromise.

The mammalian ear shows how old structures can be modified for new functions.

The body is better understood as a living museum of evolution than as a perfect divine machine.

Anatomical imperfection strongly weakens the claim that the body provides scientific evidence for God.

Key Takeaway

The body works, but it is not perfectly designed. Its weaknesses, detours, leftovers, and compromises make far more sense through evolution than through divine engineering. Evolution modifies what already exists. It does not start over. The human body is not a perfect machine from heaven. It is a historical product of natural tinkering, shaped by ancestry, survival, and time.

Sources

Charles Darwin. On the Origin of Species.

Charles Darwin. The Descent of Man.

François Jacob. Evolution and Tinkering.

Neil Shubin. Your Inner Fish.

Jerry A. Coyne. Why Evolution Is True.

Richard Dawkins. The Greatest Show on Earth.

National Academies of Sciences, Engineering, and Medicine. Science, Evolution, and Creationism.

University of California Museum of Paleontology. Understanding Evolution: Homologies and Imperfections.

E. R. Castillo and M. D. Lieberman. Lower Back Pain.

J. M. M. Oomens and colleagues. A Possible Origin of the Inverted Vertebrate Retina Revealed by Eye Ontogeny.

J. M. Culp and colleagues. Recurrent Laryngeal Nerve Injury.

Transition to the Next Chapter

The body's imperfect anatomy already weakens the claim of perfect divine design.

But anatomy is only the beginning.

The deeper challenge comes from disease.

If the body were designed by a loving and perfect creator, why is it so vulnerable to cancer, genetic disorders, infection, autoimmune disease, birth defects, pain, and aging?

The next chapter will examine disease as one of the strongest arguments against perfect design.

Chapter 12: Disease: The Body as Evidence Against Perfect Design

The Child in the Hospital Bed

Few things challenge easy faith more deeply than a sick child.

A child lies in a hospital bed, small and frightened, surrounded by tubes, monitors, medicines, alarms, and worried parents. The parents have done nothing wrong. The child has done nothing wrong. Yet the body is failing. A gene may be defective. A cancer may be growing. The immune system may be attacking its own tissues. The heart may be malformed. The blood may not clot properly. The lungs may not work as they should.

At that moment, the claim of perfect divine design becomes very hard to defend.

It is easy to praise God when the body works. It is easy to call a healthy newborn a miracle. It is easy to look at the beating heart, the seeing eye, the healing wound, and say, "This proves a designer."

But what do we say when the same body fails?

What do we say about childhood leukemia?

What do we say about cystic fibrosis?

What do we say about sickle cell disease?

What do we say about congenital heart defects?

What do we say about muscular dystrophy?

What do we say about autoimmune disease?

What do we say about cancer, stroke, sepsis, dementia, infertility, miscarriage, and chronic pain?

A fair argument must look at the whole body, not only its successes.

The human body is remarkable. But it is also vulnerable, fragile, and sometimes cruelly unfair. Disease is not an occasional scratch on a perfect machine. Disease is built deeply into the biology of life.

That is why disease is one of the strongest arguments against perfect divine design.

It makes far more sense in an evolved world than in a world designed by an all-knowing, all-powerful, all-loving God.

The Body Is Brilliant, But Breakable

The body can do astonishing things.

A cut in the skin begins to clot. White blood cells move toward infection. Fever helps fight microbes. Bones heal after fracture. The liver processes toxins. The kidneys regulate water, salt, and acid. The heart beats day and night. The lungs exchange oxygen and carbon dioxide. The brain receives the world and builds thought, memory, language, emotion, and self-awareness.

These things deserve wonder.

But wonder must be honest.

The same clotting system that saves us from bleeding can form a clot inside a blood vessel and cause a heart attack, stroke, or pulmonary embolism. The same immune system that protects us from infection can attack the thyroid, joints, skin, kidneys, nerves, intestines, or pancreas. The same cell division that allows growth and repair can become cancer. The same genetic system that carries inheritance can pass on devastating disease.

This is not perfect design.

It is biological compromise.

Evolution does not create a body with the goal of preventing all suffering. It preserves traits that help survival and reproduction well enough under certain conditions. That is a very different standard from perfect divine engineering.

The body is not designed to be flawless.

It is shaped to survive long enough, often enough, for life to continue.

That is why the body can be brilliant and breakable at the same time.

Genetic Disease: Errors in the Inheritance

A perfect designer could have made heredity safe.

There would be no fatal gene defects. No inherited blood disorders. No metabolic diseases that poison a child from within. No muscular diseases that slowly take away movement. No defective proteins that damage organs. No silent carriers passing pain to the next generation without knowing it.

But real biology is not like that.

DNA can mutate. Genes can be damaged. Chromosomes can be rearranged. Copying errors can occur. Some changes do nothing. Some cause mild disease. Some cause devastating illness. Some are inherited from parents. Some arise newly in the child. Some remain hidden for generations and then appear when two carriers have a child together.

This is not mysterious if life evolved.

Evolution depends on inheritance, but inheritance is not perfect. DNA must be copied, and copying can produce errors. Cells have repair systems, but they are not flawless. Mutations are part of the natural process of life. Without variation, evolution could not occur. But variation also brings disease.

This is the hard truth.

The same biological openness that allows evolution also allows genetic suffering.

A religious defender may say, “God has a reason.” But that does not explain the disease. It does not explain the mutation. It does not explain the pattern of inheritance. It does not help the child breathe. It does not help the family understand risk. It does not help medicine develop treatment.

Genetics does.

Genetic disease is not evidence of divine wisdom.

It is evidence of imperfect inheritance in an evolved biological system.

The Cruelty of Randomness

Disease is often unfair.

A person who never smoked may get lung cancer. A child may be born with a serious genetic disorder. A healthy young adult may develop leukemia. A pregnant woman may suffer a life-threatening complication. A person who lived carefully may develop Parkinson’s disease, multiple sclerosis, or an autoimmune condition. A newborn may arrive with a malformed heart.

Religious explanations often try to give suffering a purpose.

Some say disease is a test.

Some say it is punishment.

Some say it is part of a divine plan.

Some say God's reasons are beyond human understanding.

But these answers are not scientific. They are emotional and theological responses to pain.

Biology gives a more honest explanation.

Disease often arises from mutation, infection, inflammation, aging, developmental error, environmental exposure, inherited risk, immune dysfunction, or chance. None of these requires moral meaning. Disease does not always choose the wicked. It does not spare the kind. It does not respect innocence.

The randomness of disease fits an indifferent natural world.

It does not fit easily with the idea of a loving designer who watches every cell and knows every child.

This is not an argument from anger.

It is an argument from evidence.

A world designed by perfect love should not look so much like a world governed by blind biology.

Cancer: Growth Turned Against Life

Cancer is one of the clearest examples of the body's evolved vulnerability.

The body depends on cell division. Without cell division, a child could not grow. Skin could not renew. Blood cells could not be replaced. Wounds could not heal. The lining of the intestine could not repair itself.

But cell division carries risk.

Every time cells divide, DNA must be copied. Errors can occur. Some errors affect genes that control growth, repair, death, or restraint. If enough harmful changes accumulate, a cell may begin to grow out of control. It may invade tissues, spread through the blood or lymphatic system, and threaten the life of the whole organism.

Cancer is not an outside enemy in the same way as a bacterium.

Cancer is the body's own cellular machinery losing control.

This creates a serious problem for perfect design.

Why would a perfect designer build a body in which the very process needed for growth and repair can become a process of destruction? Why make life depend on cell division, but allow cell division to become malignant? Why make children vulnerable to leukemia, brain tumors, bone cancers, and other devastating diseases?

Evolution explains the problem more clearly.

Multicellular life requires cooperation among cells. But cooperation can break down. Natural selection shaped mechanisms to control cell growth, repair DNA, eliminate damaged cells, and protect the organism. These mechanisms are impressive, but not perfect. They reduce cancer risk. They do not remove it.

Cancer is not surprising in an evolved body.

It is deeply troubling in a perfectly designed one.

The Immune System: Protector and Attacker

The immune system is often presented as evidence of design.

It recognizes threats. It attacks bacteria and viruses. It remembers some past infections. It helps heal wounds. It defends the body every day in ways we do not even notice.

But the immune system also shows why the design argument fails.

The same system that protects us can attack us.

In autoimmune disease, the immune system mistakenly targets the body's own tissues. In rheumatoid arthritis, joints can be damaged. In type 1 diabetes, insulin-producing cells in the pancreas are destroyed. In multiple sclerosis, the nervous system is attacked. In lupus, many organs may be involved. In inflammatory bowel disease, the intestine becomes a battlefield. In autoimmune thyroid disease, the thyroid can become overactive or underactive.

Allergies and asthma show another problem. The immune system can overreact to harmless substances. A peanut, pollen, dust mite, medication, or insect sting can trigger a severe reaction. In some cases, the immune response itself can become life-threatening.

A perfect designer could create protection without self-attack.

Evolution cannot guarantee that.

The immune system must make difficult distinctions. It must respond strongly enough to danger but not so strongly that it damages the body. It must recognize countless possible threats. It must adapt. It must balance speed, power, memory, and restraint.

That balance sometimes fails.

Autoimmune disease and allergy make sense as costs of a powerful but imperfect defense system.

They do not look like perfect design.

Infection: Life Feeding on Life

Infectious disease raises another difficult question for divine design.

Why would a loving creator design a world in which living organisms survive by invading other living bodies?

Bacteria can multiply in blood, lungs, urine, wounds, brain, heart valves, and bones. Viruses enter cells and use cellular machinery to make more viruses. Fungi can infect skin, lungs, blood, or nervous tissue. Parasites can live in the intestine, blood, liver, brain, muscles, or eyes.

Some infections are mild. Others are devastating.

Smallpox killed millions. Plague reshaped societies. Tuberculosis has haunted humanity for centuries. Malaria still kills many people, especially children. HIV changed modern medicine and public health. Influenza has caused pandemics. Sepsis can turn infection into multi-organ failure.

If nature were designed by perfect love, why build a world where one form of life must invade and damage another to survive?

Evolution explains this without difficulty.

Microbes evolve. Hosts evolve. Parasites evolve. Immune systems evolve. The relationship between infection and defense is an arms race, not a peaceful design. Microbes that reproduce more successfully spread. Hosts that resist infection survive better. Each side shapes the other.

There is no need for divine planning.

There is only life competing with life.

Parasites Inside the Design Argument

Parasites are especially difficult for the idea of loving design.

Some parasites live in the intestine and steal nutrition. Some enter the blood. Some damage the liver. Some invade the brain. Some blind children. Some alter host behavior. Some pass through insects into human blood. Some worms live inside bodies for years.

A parasite is not a design error in the simple sense. It may be highly adapted to its way of life.

That is precisely the problem.

The parasite may be beautifully adapted to harm.

Its body, life cycle, reproduction, and behavior may be finely tuned to exploit another organism. If adaptation is automatically evidence of God, then parasites must also be included in God's design.

This creates a moral problem.

The same argument that praises the eye as divine design must also face the worm in the child's intestine, the parasite in the blood, and the organism that blinds or cripples its host.

Evolution explains parasitic adaptation naturally.

Parasites survive and reproduce by exploiting hosts. Traits that improve that exploitation can be favored. Hosts evolve defenses. Parasites evolve countermeasures.

No moral plan is needed.

A loving designer is much harder to defend.

Congenital Malformations: When Development Goes Wrong

Human development is astonishing.

From one fertilized cell, a body forms. Cells divide, migrate, specialize, and organize into tissues and organs. A heart forms and begins to beat. The brain and spinal cord develop. Limbs grow. Eyes, ears, kidneys, lungs, and bones take shape.

This process is remarkable.

But it is not perfect.

Development can go wrong.

A baby may be born with a malformed heart, a neural tube defect, a cleft lip or palate, kidney abnormalities, limb differences, chromosomal disorders, or many other congenital conditions. Some are mild. Some require surgery. Some cause lifelong disability. Some are fatal.

If the body were designed by a perfect creator, why would development be so vulnerable?

Evolution gives a better answer.

Development is controlled by complex genetic and cellular processes inherited from earlier life. It involves timing, signaling, growth, folding, migration, and interaction among tissues. Complex systems can fail. They are affected by genes, environment, nutrition, toxins, infections, chance, and developmental constraints.

The more complex the process, the more ways it can go wrong.

This is not surprising in evolution.

It is difficult to reconcile with perfect design.

Pregnancy and Reproduction: Life at Great Cost

Religion often speaks of reproduction as a miracle.

There is truth in the emotion. Pregnancy and birth can be deeply moving. A new child is a source of joy, hope, and love.

But reproduction also carries danger.

Miscarriage is common. Ectopic pregnancy can be life-threatening. Severe bleeding can kill. High blood pressure disorders of pregnancy can damage the mother and child. Infection can occur. Labor can become obstructed. Babies may be born too early. Some are stillborn. Some die soon after birth. Some mothers die while giving life.

A perfect designer could create reproduction without such danger.

Evolution produces reproduction through trade-offs.

The human pelvis is shaped by upright walking. The baby's head is large because of brain development. The placenta is biologically complex. The immune relationship between mother and fetus must be carefully balanced. The demands of pregnancy can strain the heart, kidneys, blood, metabolism, and blood pressure.

The system works often enough for the species to continue.

But “often enough” is not the same as perfect.

Reproduction shows the same pattern as the rest of biology: beauty mixed with risk, joy mixed with suffering, function mixed with vulnerability.

This is evolution, not flawless design.

Aging: The Body Was Not Built for Forever

Aging is another major challenge to perfect design.

The body develops, matures, reproduces, and eventually declines. Skin thins. Bones weaken. Arteries stiffen. Joints wear down. Vision changes. Hearing fades. Muscles shrink. Memory may fail. The immune system changes. Cancer risk rises. Dementia may appear. Organs lose reserve.

If the body were designed by a loving, all-powerful creator, why make aging so slow, painful, and humiliating for so many people?

Evolution offers an explanation.

Natural selection is strongest when traits affect survival and reproduction earlier in life. Harmful effects that appear late, after reproduction, may be less efficiently removed. Some traits that help early survival or reproduction may carry costs later. Repair systems are useful, but maintaining perfect repair forever may not be favored if organisms rarely survived to old age in ancestral environments.

The body is not built for immortality.

It is built for survival and reproduction under natural constraints.

This does not make aging less painful. It makes it more understandable.

Aging is not evidence of divine perfection.

It is evidence of biological limits.

Pain: Protection That Can Become Torture

Pain is useful.

Without pain, we might not pull our hand away from fire. We might walk on a broken bone. We might ignore infection, injury, pressure, or internal damage. Pain warns, protects, and teaches.

But pain can also become terrible.

It can persist after tissues heal. It can become chronic. It can arise from damaged nerves. It can torment patients with cancer, arthritis, spinal disease, neuropathy, migraines, autoimmune disease, sickle cell crises, and many other conditions.

A system designed to protect can become a system that tortures.

This is another example of biological compromise.

Evolution can explain pain as a protective signal that sometimes misfires, persists, or becomes maladaptive. It does not need to be perfect. It only needs to provide enough survival advantage to persist.

But perfect divine design faces a harder question.

Why make pain so severe? Why make it chronic? Why allow infants, children, animals, and innocent people to suffer without understanding or relief? Why make a protective system capable of becoming unbearable?

Pain shows that the body is not arranged around mercy.

It is arranged around survival.

Mental Illness and the Vulnerable Brain

The brain is often treated as the crown of creation.

It allows thought, memory, language, imagination, love, planning, music, morality, science, and self-awareness. It is an astonishing organ.

But the brain is also vulnerable.

It can develop depression, anxiety, psychosis, addiction, obsessive thoughts, dementia, seizures, trauma responses, mood disorders, and many other conditions. Some people suffer silently for years. Some lose their sense of reality. Some lose memory. Some lose the ability to care for themselves. Some lose the desire to live.

If the brain were divinely designed as the seat of a soul, why is it so easily disturbed by chemistry, injury, genes, inflammation, drugs, sleep loss, trauma, degeneration, and disease?

A natural view makes more sense.

The brain is an evolved organ. It is powerful, but fragile. It developed through layers of older systems and newer abilities. It must balance emotion, memory, threat detection, reward, social connection, attention, and body regulation. These systems can help survival, but they can also fail.

Mental illness is not demon possession. It is not spiritual weakness. It is not divine punishment.

It is biology, psychology, environment, and experience interacting in a vulnerable brain.

This truth has humane power.

It replaces superstition with care.

Evolutionary Trade-Offs

Many diseases make more sense when we understand trade-offs.

A trait that helps in one way may harm in another. A strong immune response may fight infection but increase inflammatory damage. Blood clotting prevents bleeding but can cause dangerous clots. High stomach acid helps digestion and defense but can contribute to ulcers or reflux symptoms. Cell division supports growth and repair but increases cancer risk. Anxiety can help detect danger but can become disabling in modern life.

Evolution does not eliminate all trade-offs.

It preserves what works well enough in a specific environment.

This means disease is not always a simple defect. Sometimes it is the cost of a system that also has benefits. Sometimes it is a mismatch between bodies shaped in earlier environments and modern conditions. Sometimes it is the result of aging, mutation, infection, or chance.

Perfect design should avoid cruel trade-offs.

Evolution is built from them.

This is why evolutionary medicine is so important. It does not only ask how disease happens. It asks why bodies are vulnerable to disease in the first place.

That question points away from perfect design and toward natural history.

Modern Mismatch: Old Bodies in a New World

Many modern diseases are influenced by mismatch.

Our bodies evolved in environments very different from modern life. For much of human history, people had different diets, more physical activity, more sunlight exposure, different sleep patterns, more infections, shorter life expectancy, and different social pressures.

Today many people live with processed food, high calorie intake, low physical activity, artificial light, chronic stress, social isolation, pollution, tobacco, alcohol, drugs, and long lifespans. Bodies shaped by earlier environments now face new conditions.

This mismatch can contribute to obesity, type 2 diabetes, high blood pressure, heart disease, sleep problems, anxiety, depression, musculoskeletal pain, and many other conditions.

This is not because the body is evil.

It is because the body is historical.

It was not designed for every possible future. It was shaped by past environments. When environments change faster than biology can adapt, disease risk can rise.

A perfect designer could anticipate every future.

Evolution cannot.

Why Disease Is Not a Small Problem for Design

Some religious believers say disease entered the world because of sin, disobedience, or a fall from grace.

This explanation does not work scientifically.

Disease existed long before modern humans. Fossils show signs of injury, infection, tumors, arthritis, and parasitism in ancient life. Animals suffered before humans wrote scriptures, built temples, or developed moral codes. Predation, infection, birth defects, and death were part of life long before any human religious story was told.

So disease cannot honestly be explained as the result of human sin in a biological sense.

It is older than humanity.

This is a serious problem for the claim that suffering came from human disobedience. Nature was already filled with suffering before humans existed.

Evolution explains that naturally.

Life has always involved mutation, competition, infection, injury, reproduction, aging, and death. Disease is part of the biological fabric, not a late moral punishment.

The design argument often avoids this fact.

But science cannot avoid it.

The Failure of "Mysterious Plan"

When faced with disease, many religious people say, "God has a mysterious plan."

This may comfort some people, but it is not an explanation.

A mysterious plan can explain anything. A healthy child is God's plan. A dying child is God's plan. Recovery is God's plan. Death is God's plan. A cure is God's plan. No cure is God's plan.

If every outcome is called God's plan, then the phrase has no scientific meaning.

Medicine cannot work that way.

A physician cannot treat pneumonia by saying it is mysterious. A surgeon cannot repair an appendix by saying it is God's will. A genetic counselor cannot help a family by saying the mutation is part of a divine plan. Public health cannot prevent infection by appealing to mystery.

Science works because it looks for real causes.

Bacteria. Viruses. Mutations. Inflammation. Tumors. Clots. Hormones. Toxins. Deficiencies. Trauma. Developmental errors. Aging. Environment.

Once causes are understood, suffering can often be reduced.

The mysterious plan explains nothing and prevents nothing.

Natural explanation gives us a path.

Medicine Is a Quiet Refutation of Superstition

Every hospital is, in one sense, a monument to natural explanation.

When a patient arrives with chest pain, doctors do not ask which demon entered the heart. They check the EKG, cardiac enzymes, arteries, oxygen, blood pressure, rhythm, and risk factors. When a child has meningitis, doctors do not search for curses. They look for infection and treat urgently. When a tumor is found, doctors do not assume divine punishment. They biopsy, stage, operate, irradiate, or treat with chemotherapy, immunotherapy, targeted therapy, or palliation.

Medicine works because disease has natural causes.

That fact matters for this book.

If disease is natural, then the religious interpretation of disease as punishment, test, curse, or divine plan loses scientific authority. Faith may comfort some patients emotionally, but it does not explain the disease process.

Science explains.

Science does not always cure. It does not always save. It does not always have a perfect answer. But it asks the right kind of question.

What is happening in the body?

What caused it?

How can it be prevented?

How can it be treated?

How can suffering be reduced?

These are humane questions.

They do more for human beings than superstition ever did.

The Body as Evidence for Evolution

Disease makes sense in evolution because evolution predicts imperfection.

DNA mutates.

Cells make errors.

Immune systems misfire.

Microbes adapt.

Parasites exploit.

Organs wear down.

Development can fail.

Pain can persist.

Aging weakens the body.

Trade-offs create vulnerability.

Old bodies meet new environments.

This is exactly the kind of world evolution would produce.

It is not exactly the kind of world perfect design would lead us to expect.

A believer may still say God is hidden behind the process. But that claim adds nothing scientific. It does not explain why the genome mutates. It does not explain cancer. It does not explain autoimmunity. It does not explain sepsis. It does not explain childhood disease. It does not explain aging.

Evolutionary biology and medicine do explain these things.

Not completely in every detail, but increasingly and usefully.

That is the difference.

Science grows.

God of the gaps retreats.

A Human Response Without God

Some people worry that without God, suffering becomes meaningless.

But calling suffering God's plan does not make it more meaningful. In many cases, it makes it more morally troubling. If God planned or allowed every disease for a higher purpose, then God becomes responsible for enormous suffering that no decent human being would willingly design.

Science-based atheism offers a different response.

Disease is not a divine message.

Disease is not punishment.

Disease is not a sacred test.

Disease is not evidence of a hidden plan.

Disease is a natural reality.

That may sound hard, but it can also be deeply compassionate. It removes guilt. It removes superstition. It removes the cruel idea that illness reflects sin, weak faith, karma, or divine displeasure.

A child with cancer is not being tested by God.

A mother with autoimmune disease is not being punished.

An old man with dementia is not spiritually cursed.

They are human beings caught in the vulnerability of biological life.

Our response should not be worshipful explanation.

It should be care.

The Calm Conclusion

Disease is one of the strongest arguments against perfect divine design.

The body is wonderful, but it is not flawless. The genome is powerful, but it mutates. Cells grow, but they can become cancer. The immune system protects, but it can attack the body. Reproduction creates life, but it can bring miscarriage, birth defects, and maternal death. The brain allows consciousness, but it can suffer mental illness and dementia. Pain protects, but it can become torture. Aging brings decline. Infection reveals life feeding on life.

This pattern is not accidental to biology.

It is part of biology.

Evolution explains why the body is vulnerable. It explains trade-offs, mutation, selection, mismatch, infection, aging, and imperfect repair. It explains why life can be extraordinary and fragile at the same time.

Perfect divine design does not explain this well.

If the human body were made by a loving, all-powerful, all-knowing designer, disease would be one of the hardest facts to understand. But if the body is the result of evolution, disease becomes tragic but expected.

The conclusion is clear:

The body does not provide scientific evidence of perfect design.

Disease reveals the truth.

Life is evolved, fragile, and natural.

No divine designer is required.

Chapter Summary

Disease is one of the strongest challenges to the idea of perfect divine design.

The body is remarkable, but it is also fragile and vulnerable.

Genetic disease arises because inheritance is powerful but imperfect.

Cancer shows how cell division, necessary for growth and repair, can become destructive.

The immune system protects the body, but it can also cause autoimmune disease, allergy, and inflammation.

Infectious disease shows life exploiting life through bacteria, viruses, fungi, and parasites.

Congenital malformations reveal the vulnerability of complex development.

Pregnancy, childbirth, pain, aging, and mental illness all show biological trade-offs and limits.

Disease existed long before humans, so it cannot be explained scientifically as the result of human sin.

Medicine works because disease has natural causes, not supernatural ones.

Evolution explains disease vulnerability far better than perfect divine design.

Key Takeaway

Disease exposes the weakness of the design argument. The body is not a flawless divine machine. It is an evolved system shaped by mutation, trade-offs, infection, aging, imperfect repair, and biological limits. Disease is painful and often unfair, but it is not mysterious evidence of God's plan. It is evidence that life is natural, fragile, and evolved.

Sources

Randolph M. Nesse and George C. Williams. Why We Get Sick: The New Science of Darwinian Medicine.

Randolph M. Nesse. Good Reasons for Bad Feelings.

Peter D. Gluckman, Alan Beedle, and Mark Hanson. Principles of Evolutionary Medicine.

Stephen C. Stearns and Jacob C. Koella. Evolution in Health and Disease.

M. L. Benton and colleagues. The Influence of Evolutionary History on Human Health and Disease.

National Human Genome Research Institute. Genetic Disorders.

National Cancer Institute. What Is Cancer?

Centers for Disease Control and Prevention. About Antimicrobial Resistance.

Centers for Disease Control and Prevention. Antimicrobial Resistance: Causes and How It Spreads.

World Health Organization. Antimicrobial Resistance and Infectious Disease Resources.

Transition to the Next Chapter

Disease already shows that life is not perfectly designed.

But the problem goes beyond the human body.

Nature itself is filled with suffering, predation, parasitism, starvation, waste, injury, and death. Much of this suffering existed long before humans appeared.

The next chapter will look beyond the hospital and into the wider natural world, where life survives by consuming, invading, competing with, and destroying other life.

Chapter 13: Nature Without Mercy: Parasites, Predation, and Waste

The Wasp That Troubled Darwin

Charles Darwin was not troubled only by fossils, anatomy, and variation.

He was also troubled by cruelty in nature.

One example disturbed him deeply: parasitic wasps.

Certain wasps lay their eggs inside the bodies of living caterpillars or other hosts. The young wasps develop inside the host, feeding on it from within. The host may remain alive for some time while its body is slowly consumed.

This is not an accident.

It is a way of life.

To a person who believes nature clearly reveals the kindness of God, such examples create a serious problem. A flower in bloom may inspire prayer. A bird singing at dawn may feel like a hymn. But what about the larva eating a living body from the inside? What about the parasite that blinds a child? What about the virus that enters human cells and uses them to make more virus?

What about the predator tearing open its prey while the prey is still alive?

If nature is designed by perfect love, why is so much of it built on suffering?

This chapter looks beyond the hospital and into the wider living world. Disease already challenges divine design. But nature itself creates an even broader challenge. The natural world is filled with predation, parasitism, starvation, injury, competition, reproductive waste, and death.

Much of this suffering existed long before human beings appeared.

Before scripture.

Before sin.

Before temples.

Before prayer.

Before any human being could disobey any god.

This is not a world designed around mercy.

It is a world shaped by survival.

The Romantic Lie About Nature

Human beings often romanticize nature.

We admire mountains, forests, rivers, birds, flowers, and sunsets. We speak of the balance of nature. We describe nature as wise, healing, peaceful, and pure. Religious traditions often go further and treat nature as evidence of divine goodness.

There is truth in the beauty.

A forest can be beautiful. A bird in flight can be beautiful. The ocean can be magnificent. A mother animal caring for her young can touch the heart. The natural world can quiet the mind and awaken deep wonder.

But beauty is not the whole truth.

The same forest contains hunger, fear, parasites, broken bones, infected wounds, abandoned young, territorial violence, and slow death. The same ocean contains predation, suffocation, disease, and creatures consumed alive. The same meadow that holds flowers also holds insects eating other insects, fungi spreading through living tissue, and animals dying unseen in the grass.

Nature is not a moral teacher.

Nature is a field of reproduction, survival, competition, cooperation, and death.

This does not mean nature is evil. Evil requires intention. Nature has no intention. It does not plan suffering. It does not enjoy cruelty. It does not hate the weak.

Nature is not cruel in a human moral sense.

Nature is indifferent.

That indifference is exactly what evolution predicts.

It is not what we would expect from a loving divine designer.

Life Feeds on Life

One of the hardest facts about nature is also one of the simplest:

Life feeds on life.

A lion must kill to live. A hawk must seize prey. A snake swallows another animal whole. A spider traps insects. A fish eats smaller fish. A bacterium invades tissue. A virus enters cells. A fungus digests living or dead material. A parasite takes nourishment from its host.

This is not an exception to nature.

It is central to nature.

Religious descriptions of creation often soften this reality. They speak of beauty, order, and harmony. But much of the natural world is not harmonious from the viewpoint of the individual animal being hunted, infected, starved, or consumed.

The gazelle does not experience the lion as harmony.

The caterpillar does not experience the wasp larva as divine balance.

The child with malaria does not experience the parasite as evidence of heavenly wisdom.

From the standpoint of evolution, this is not surprising. Organisms compete for energy, space, mates, safety, and reproduction. Some survive by eating plants. Some survive by eating animals. Some survive by invading other bodies. Some survive by manipulating host behavior. Some survive by producing vast numbers of offspring, most of which will die.

Evolution has no moral preference for kindness.

It preserves what works.

If harming another organism helps survival and reproduction, that strategy can spread.

A loving designer is not needed to explain this.

A loving designer is made harder to defend by it.

Predation: Beauty With Terror Inside

Predators can be magnificent.

A tiger moving through tall grass is beautiful. An eagle diving through the air is beautiful. A shark moving through water is powerful and elegant. A wolf pack hunting together shows intelligence, coordination, and endurance.

But predation is not beautiful to the prey.

To the prey, it is terror.

The deer feels the chase. The rabbit feels the jaws. The fish feels the bite. The insect struggles in the web. The young animal may be taken before it has lived more than a few days or weeks.

Predation shapes bodies and behavior. It creates speed, armor, venom, claws, teeth, camouflage, alertness, herd behavior, hiding, and flight. Predators become better hunters. Prey become better escapers. This struggle can produce remarkable adaptations.

But adaptation is not mercy.

The existence of an efficient predator does not prove divine design. It proves that organisms that hunt effectively can survive and reproduce. The existence of fast prey does not prove that God designed a balanced system. It shows that prey unable to escape were more likely to die before reproducing.

Predation is evolution in motion.

It is not a moral arrangement from heaven.

If a human being created a world where one group of creatures must terrify, injure, and tear apart another group in order to live, we would not call that person morally perfect. Yet when nature works this way, religion often calls it divine wisdom.

Science gives a simpler answer.

There is no moral designer behind it.

Parasites: Adapted to Harm

Parasites create one of the strongest problems for divine design.

A parasite is often beautifully adapted to its harmful way of life.

Some parasites attach to the intestine and steal nutrients. Some live in blood. Some invade organs. Some pass through insects into human bodies. Some manipulate the behavior of hosts. Some damage eyes, brains, liver, lungs, skin, or muscles. Some cause chronic illness. Some kill.

The disturbing point is not merely that parasites exist.

The disturbing point is that many are highly adapted.

Their life cycles can be complex. Their bodies may be specialized. Their timing can be precise. Their ability to evade the immune system can be impressive. Their reproductive strategies can be efficient.

If adaptation proves God, then parasites also prove God.

That is the problem.

The same argument that praises the hummingbird must also explain the tapeworm. The same argument that praises the eye must also explain the parasite that blinds the eye. The same argument that praises the immune system must also explain the organism that evades it.

Evolution explains parasites without moral confusion.

Parasites survive by exploiting hosts. Traits that improve exploitation can be favored. Hosts evolve defenses.

Parasites evolve ways around those defenses. The result is an arms race, not a divine plan.

No loving designer is required.

In fact, a loving designer becomes less believable.

The Parasitoid Wasp and the Collapse of Easy Faith

The parasitoid wasp deserves special attention because it exposes the weakness of sentimental religion.

It is easy to see God in a flower.

It is harder to see God in a wasp larva consuming a living caterpillar from within.

Darwin understood this problem. He saw that nature did not simply display beauty and harmony. It also displayed suffering built into reproductive strategy. The wasp did not accidentally harm the caterpillar. The wasp's life cycle depended on it.

This is why the problem is so deep.

A parasite that harms by accident would be one kind of problem. A parasite whose entire reproductive success depends on prolonged harm is another. The suffering is not incidental. It is part of the system.

The religious mind may say, "God's ways are mysterious."

But mystery is not explanation.

Evolution explains the wasp clearly. The wasp's ancestors that used hosts more successfully left more descendants. Traits that improved egg laying, host selection, larval survival, and reproductive success were favored. The process did not require pity. It required only survival and reproduction.

The wasp does not disprove every possible god by itself.

But it destroys the easy claim that nature plainly reveals a loving designer.

Nature reveals adaptation.

Adaptation can be horrifying.

Starvation: The Quiet Suffering of Nature

Not all suffering in nature is dramatic.

Some of it is quiet.

A young animal cannot find food. A bird with an injured wing slowly weakens. A deer in winter uses its last energy searching through snow. A seal pup loses its mother. A drought dries the plants. A fish trapped in shrinking water suffocates. A predator too old to hunt starves. A weak newborn cannot compete with stronger siblings.

There may be no roar, no blood, no visible attack.

Only hunger.

Starvation is common in the natural world because resources are limited. Many organisms produce more offspring than the environment can support. This is one reason natural selection works. Not all survive. Those with traits better suited to the conditions are more likely to reproduce. Others die.

This is biologically understandable.

But it is not morally beautiful.

A loving designer could create life without such waste. A loving designer could create a world where survival does not depend on scarcity, hunger, and the death of the weak. Evolution cannot do that. Evolution operates within competition and limitation.

Starvation does not look like divine care.

It looks like nature without mercy.

Wasteful Reproduction

Many forms of life reproduce in enormous numbers because most offspring will die.

A fish may release thousands or millions of eggs. A sea turtle may lay many eggs, but only a small number of hatchlings survive to adulthood. Insects may produce large numbers of offspring. Plants may release countless seeds, most of which never become mature plants.

From the perspective of evolution, this makes sense.

If many offspring die, producing more can increase the chance that some survive. Natural selection does not care about the suffering of the many that perish. It only preserves reproductive strategies that work over generations.

But from the perspective of perfect design, this is strange.

Why create a system that depends on enormous waste? Why make life continue through the mass death of offspring? Why make survival a lottery in which most young never reach maturity?

Some people call this the balance of nature.

But balance is a soft word for a harsh reality.

Millions are born so that a few may live.

This is not the work of a careful moral planner.

It is the result of reproduction under natural selection.

Sibling Competition and the Struggle Within the Nest

Nature is not always gentle even within the family.

In some species, siblings compete intensely for food. A stronger chick may push a weaker one aside. In some birds, one chick may kill another. In some mammals, stronger offspring may receive more nourishment while weaker ones decline. In harsh conditions, parents may feed the young most likely to survive.

This can be painful to observe.

We often imagine parental care as pure love, but nature is more complicated. Parental behavior is also shaped by survival and reproductive success. In some environments, investing limited resources in the strongest offspring may lead to more surviving descendants.

Again, this makes evolutionary sense.

But it is not a moral ideal.

If nature were a lesson written by a loving God, why would sibling rivalry become lethal? Why would the weak be pushed aside? Why would parental care sometimes become selective in such a hard way?

Evolution answers without needing moral comfort.

Organisms behave in ways shaped by reproductive pressures, resource limits, and inherited tendencies.

There is no divine kindness guiding every nest.

Infanticide and Reproductive Strategy

Some animal behavior is deeply disturbing to human moral intuition.

In certain species, males may kill young offspring that are not their own, making the female more likely to become fertile again. In other settings, offspring may be abandoned, consumed, or neglected under harsh conditions. Not every species behaves this way, and not every act has the same explanation, but such behaviors exist in nature.

These facts are difficult for religious romanticism.

They show that reproduction in nature is not organized around human ideals of love, justice, or innocence. It is shaped by reproductive success, competition, environment, and survival.

Evolution does not say these behaviors are morally good.

That distinction is crucial.

To explain a behavior is not to approve of it. Science can explain infanticide, aggression, parasitism, or predation without calling them good. Evolution describes what happens and why certain traits or behaviors may spread. It does not provide moral approval.

This is another reason science-based atheism is often more honest than religious design.

It does not pretend that nature is morally arranged.

It admits that morality must come from human reflection, compassion, reason, and social life, not from copying nature.

Sexual Selection and Harm

Sexual selection can produce beauty.

Bird songs, bright colors, courtship dances, antlers, displays, and ornaments can be wonderful to observe. Nature can seem artistic.

But sexual selection can also produce harm.

Competition for mates can lead to violence. Males may fight and injure each other. Some mating behaviors can harm females. Some traits that improve mating success may reduce survival. Bright colors may attract predators. Large antlers may cost energy. Displays may expose animals to danger.

Beauty and burden often come together.

This is not perfect design.

It is reproductive competition.

A trait may spread if it helps mating success, even if it carries costs. Natural selection and sexual selection do not aim at happiness. They do not aim at kindness. They do not aim at the best life for each individual. They shape traits that influence survival and reproduction.

This is why nature can be beautiful and harsh at the same time.

The peacock's tail may be beautiful, but beauty alone is not evidence of God.

It is also evidence of selection.

Animal Suffering Before Humans

This point is essential:

Animal suffering existed long before humans.

Predation existed before human beings. Parasites existed before human beings. Disease existed before human beings. Injury, starvation, birth defects, competition, and death existed before human beings. Fossils show a long history of life, struggle, trauma, and extinction before our species appeared.

This creates a serious problem for religious explanations that blame suffering on human sin.

If animals were suffering millions of years before humans existed, then suffering cannot be explained scientifically as the result of human disobedience. Dinosaurs were injured and diseased before any human prayer was spoken. Parasites existed before temples. Predators hunted before scripture. Extinction occurred before morality.

The natural world was not a peaceful paradise ruined by humans.

It was already filled with suffering.

This does not mean humans cannot add suffering. We certainly can, and we often do. But biological suffering is older than humanity.

Evolution explains this naturally.

Religion struggles with it.

The Problem of Natural Evil

Philosophers often distinguish between moral evil and natural evil.

Moral evil includes suffering caused by human actions: murder, cruelty, war, abuse, oppression, and injustice.

Natural evil includes suffering caused by nature: disease, earthquakes, floods, predation, parasites, genetic disorders, starvation, and animal pain.

This book focuses on the biological side of natural evil.

If a human being harms another human being, religion may appeal to free will. That argument has its own problems, but at least it tries to explain human-caused harm.

But free will does not explain malaria.

Free will does not explain childhood cancer.

Free will does not explain parasitic wasps.

Free will does not explain a gazelle being eaten alive.

Free will does not explain animals starving before humans existed.

Natural evil is a deep problem for the idea of a loving and powerful God.

Evolution makes sense of it because evolution has no moral intention. It does not promise fairness. It does not protect innocence. It does not prevent suffering. It does not create paradise.

It creates life that continues through reproduction, variation, competition, cooperation, and death.

That is why science explains nature better than theology.

Why Nature Is Not a Moral Guide

Some people say we should live according to nature.

This can be dangerous.

Nature contains cooperation, care, and beauty, but it also contains exploitation, disease, violence, and death. If we copy nature uncritically, we can justify almost anything.

Predation is natural. That does not make human violence moral.

Infanticide occurs in nature. That does not make it acceptable.

Dominance hierarchies occur in nature. That does not justify oppression.

Parasites survive naturally. That does not make exploitation good.

Science tells us what nature is.

It does not tell us that everything natural is morally right.

This is another place where science-based atheism must be mature. If there is no God, morality does not come from copying nature. It comes from human needs, suffering, reason, empathy, cooperation, and the recognition that conscious beings can be harmed or helped.

Nature is not our moral master.

Nature is our origin.

We can learn from it, but we should not worship it.

Cooperation Exists, But It Does Not Rescue Design

A fair discussion must acknowledge that nature is not only violent.

There is cooperation in nature. Parents care for young. Social animals live in groups. Some species warn one another of danger. Mutualistic relationships exist between plants and pollinators, fungi and plants, gut bacteria and animals, and many other forms of life. Cooperation can be powerful.

But cooperation does not rescue divine design.

Evolution can explain cooperation too.

Cooperation can evolve when it benefits survival or reproduction, especially among relatives, reciprocal partners, social groups, or organisms whose interests overlap. Mutual benefit can shape behavior just as competition does.

This is important because the argument is not that nature is only cruel. The argument is that nature is morally indifferent.

Nature contains care and harm, cooperation and exploitation, beauty and terror.

That mixture is exactly what evolution predicts. It is not what we would expect from a perfect moral designer whose nature is pure love.

The existence of cooperation does not prove God any more than the existence of parasitism disproves every possible god.

The whole pattern matters.

The pattern is not divine love.

The pattern is natural selection, kinship, competition, mutual benefit, conflict, and survival.

"Balance of Nature" Is Not Moral Balance

The phrase "balance of nature" can be misleading.

Ecosystems do have relationships. Predators affect prey. Plants affect herbivores. Pollinators affect flowers. Microbes affect soil. Changes in one species can affect others. There are cycles, dependencies, and feedback systems.

But this balance is not moral balance.

A stable ecosystem can still contain immense suffering. A parasite can be part of an ecosystem. A predator-prey relationship can be ecologically important and still terrifying for the prey. Disease can regulate populations. Starvation can limit numbers. Death can feed other life.

Calling this balance does not make it kind.

From the perspective of the system, death may recycle nutrients. From the perspective of the dying individual, it is suffering.

Religion often zooms out and calls nature wise.

Science forces us to look at both the system and the individual.

Evolution explains the system without pretending it is merciful.

The Indifference of Nature and the Responsibility of Humans

If nature is indifferent, does that lead to despair?

It should not.

It should lead to responsibility.

If there is no divine mercy built into nature, then human mercy matters more. If no god prevents suffering, then medicine matters more. If no cosmic justice protects the weak, then human justice matters more. If animals can suffer, then compassion matters more. If children are not protected by supernatural design, then science, care, and prevention matter more.

Atheism does not require coldness.

It can create a deeper moral urgency.

Because nature does not care, we must care.

Because evolution is blind, we must see.

Because there is no divine plan ensuring goodness, we must build goodness.

This is one of the most humane conclusions of science-based atheism. It does not deny suffering. It does not explain it away as mysterious love. It does not call pain sacred. It says suffering is real, and conscious beings deserve compassion.

That is more honest than calling every cruelty part of God's plan.

Why Divine Design Fails Here

The design argument usually selects the beautiful parts of nature.

It points to flowers, wings, eyes, sunsets, and newborn babies.

But a serious argument must include everything.

If the flower is evidence of God, what is the parasite?

If the bird song is evidence of God, what is the animal screaming in a predator's mouth?

If the healthy newborn is evidence of God, what is the newborn dying of infection?

If the harmony of nature is evidence of God, what is mass starvation?

If adaptation proves design, does parasitic adaptation also prove design?

The religious answer often becomes selective.

Beauty proves God.

Cruelty is mystery.

Health proves God.

Disease is mystery.

Survival proves God.

Extinction is mystery.

This is not honest reasoning.

Evolution does not need to select only pleasant facts. It can explain beauty and brutality in the same framework. It explains flowers and parasites, cooperation and competition, care and predation, survival and extinction.

That is why evolution is intellectually stronger.

It explains the whole living world, not only the parts we like.

Nature Without a Loving Author

A book has an author.

A machine has an engineer.

A painting has a painter.

But nature does not behave like a moral work written by a loving author.

If nature were a book written by perfect love, it would contain many brutal chapters. If nature were a machine designed by perfect wisdom, it would contain much unnecessary suffering. If nature were a painting of divine goodness, it would be painted with blood, hunger, infection, and extinction as well as flowers, birds, and light.

A more honest conclusion is that nature has no author in that sense.

Nature is not written.

Nature unfolds.

It unfolds through physics, chemistry, biology, evolution, ecology, and time. Life emerges, changes, competes, cooperates, suffers, reproduces, and dies. Some forms persist. Many vanish. There is no guarantee of fairness. There is no promise of mercy.

This may be emotionally difficult, but it fits the evidence.

And truth should not be rejected because it is uncomfortable.

Wonder Without Illusion

Seeing nature clearly does not require losing wonder.

A predator is still powerful. A bird is still beautiful. A forest is still calming. A whale is still majestic. A flower is still delicate. A spider web is still remarkable. The living world is still astonishing.

But our wonder must mature.

Childlike wonder says, “It is beautiful, so God made it.”

Mature wonder says, “It is beautiful, and I want to understand how nature produced it.”

Mature wonder does not close its eyes to suffering. It does not use flowers to hide parasites. It does not use sunsets to erase starvation. It does not use birdsong to silence the cry of prey.

Science allows us to love nature without lying about it.

That is a better kind of wonder.

It is wonder without worship.

The Calm Conclusion

Nature is beautiful, but it is not merciful.

It contains cooperation, care, and astonishing adaptation. But it also contains predation, parasites, starvation, disease, injury, reproductive waste, sibling competition, infanticide, sexual conflict, and immense animal suffering.

This is not a small problem for divine design.

It is central.

A world created by perfect love should not look so much like a world shaped by blind survival. A world designed by moral wisdom should not require so much fear, hunger, invasion, and death. A world made for divine goodness should not contain parasites adapted to torment, predators adapted to kill, and offspring produced in vast numbers so that most can die.

Evolution explains this world.

It explains beauty and brutality through the same natural processes. It explains adaptation without kindness. It explains cooperation without divine love. It explains suffering without sacred purpose. It explains why life survives by using other life.

Nature is not cruel because it has evil intentions.

Nature is indifferent because it has no intentions.

That is the difficult truth.

But it is also the honest truth.

The living world does not show scientific evidence of a loving designer.

It shows evolution, struggle, reproduction, selection, and death.

No divine mercy is built into nature.

No divine designer is required.

Chapter Summary

Nature is beautiful, but it is also filled with suffering, predation, parasitism, starvation, injury, and death.

The romantic view of nature often ignores the pain built into natural life.

Life feeds on life, and many organisms survive by killing, invading, or exploiting others.

Predation creates powerful adaptations, but it also creates terror and suffering for prey.

Parasites are especially difficult for divine design because many are highly adapted to harm.

Wasteful reproduction means many organisms are born so that only a few survive.

Animal suffering existed long before humans, so it cannot be explained scientifically as the result of human sin.

Natural evil is a major problem for belief in a loving and powerful God.

Nature contains cooperation as well as harm, but this mixture is better explained by evolution than by perfect moral design.

Science-based atheism does not make us less compassionate. Because nature is indifferent, human compassion becomes more important.

Key Takeaway

Nature does not behave like the creation of a loving moral designer. It contains beauty and care, but also parasites, predation, starvation, waste, and immense suffering. This pattern is exactly what evolution predicts: life shaped by survival, reproduction, competition, cooperation, and chance. Nature is not cruel by intention. It is indifferent because it has no intention. The living world does not reveal divine mercy. It reveals natural history.

Sources

Charles Darwin. On the Origin of Species.

Charles Darwin. Letter to Asa Gray, May 22, 1860.

Richard Dawkins. The Greatest Show on Earth.

Jerry A. Coyne. Why Evolution Is True.

David Attenborough. Life on Earth.

George C. Williams. Adaptation and Natural Selection.

Randolph M. Nesse and George C. Williams. Why We Get Sick.

University of California Museum of Paleontology. Understanding Evolution: Coevolution.

University of California Museum of Paleontology. Understanding Evolution: Natural Selection.

National Academies of Sciences, Engineering, and Medicine. Teaching About Evolution and the Nature of Science.

Transition to the Next Chapter

Nature's suffering already weakens the claim that life was designed by perfect love.

But the history of life contains an even larger challenge.

Most species that ever lived are gone.

Entire worlds of life appeared, flourished, struggled, and vanished. The fossil record is not only a record of life. It is also a record of disappearance.

The next chapter will examine extinction, the graveyard of divine design.

Chapter 14: Extinction: The Graveyard of Divine Design

The World That Disappeared

Stand in a museum before the skeleton of a dinosaur.

The bones rise above you like the frame of a vanished world. The jaws are huge. The ribs curve like the beams of an ancient ship. The legs look powerful enough to shake the ground. Children stare upward in amazement. Adults feel the same wonder, though they may hide it better.

But behind the wonder is a harder truth.

This creature is gone.

Not merely dead.

Extinct.

Its world is gone too. The plants it walked past, the air it breathed, the enemies it feared, the young it protected, the landscapes it knew, all belong to a chapter of life that ended long before human beings appeared.

The dinosaur is not alone.

The history of life is filled with vanished forms: trilobites, ammonites, giant marine reptiles, mammoths, saber-toothed cats, ancient birds, early whales, strange mammals, and many human relatives. Some lived for millions of years. Some were successful in their time. Some were beautiful, powerful, complex, and well adapted.

Then they disappeared.

If life were the work of a perfect divine designer, extinction creates a serious problem. Why would an all-knowing creator make countless species, allow them to struggle, reproduce, suffer, adapt, and then erase most of them? Why create entire worlds of life only to bury them in rock?

The fossil record is not a hymn to perfect design.

It is a graveyard.

Most Life Is Gone

The living world around us feels abundant.

There are birds in the sky, insects in the grass, fish in the sea, trees in forests, fungi in soil, bacteria on every surface, animals in deserts, plants on mountains, and human beings in cities across the planet. Life seems everywhere.

But what we see today is only the surviving edge of a much larger history.

Most species that ever lived are extinct.

This fact is difficult to absorb. The present world is not the full story of life. It is only the remaining page of a long book, most of whose characters are gone.

Extinction is not a rare accident. It is part of life's history. Species appear, persist for a time, and disappear. Some vanish quietly as environments change. Some are outcompeted. Some lose habitat. Some are struck by disease. Some fail when climate shifts. Some die in great mass extinctions that erase large portions of life in a relatively short geological time.

This is not what many people imagine when they speak of divine creation.

Creation stories often present life as purposeful, ordered, and complete. But the fossil record shows a restless and unstable history. Life does not sit peacefully in fixed categories. It branches, changes, struggles, and dies.

Extinction is one of the clearest signs that the living world is not a perfect plan.

It is natural history.

Extinction Is Not Failure in Evolution

From a human emotional point of view, extinction feels like failure.

A species disappears forever. No more songs. No more footprints. No more nests. No more young. No more descendants. A whole way of being alive comes to an end.

But in evolution, extinction is not a violation of the process.

It is part of the process.

Evolution does not promise that every species will survive. It does not protect complexity. It does not reward beauty. It does not preserve intelligence. It does not maintain every branch because it once flourished.

A species survives only as long as its inherited traits remain good enough in changing conditions.

When conditions change too much, a species may disappear.

A body that worked well in one climate may fail in another. A predator that hunted well in one ecosystem may starve when prey vanish. A plant adapted to one pattern of rainfall may fail when the climate shifts. A large animal may need more food than a damaged environment can provide. A specialized organism may lose the one niche on which it depends.

Evolution has no loyalty.

This is why extinction fits evolution so well.

It does not fit perfect divine care nearly as well.

Dead Branches on the Tree of Life

The tree of life is not a neat tree where every branch continues upward.

It is full of dead branches.

Some branches split and flourish. Some branch again. Some shrink. Some end. The living species we see today are only the tips of the branches that survived to the present.

This matters because many people imagine evolution as a straight line of progress. They picture simple life becoming fish, fish becoming reptiles, reptiles becoming mammals, mammals becoming monkeys, and monkeys becoming humans.

That picture is wrong.

Evolution is not a ladder.

It is a branching tree, and most branches have ended.

Dinosaurs were not failed mammals. Trilobites were not waiting to become something better. Neanderthals were not incomplete humans. Each lineage had its own history. Each lived in its own conditions. Each survived for a time. Then many disappeared.

A divine plan centered on humans would not naturally predict such wasteful branching.

But evolution does.

Branches form because populations separate and change. Branches end because extinction is common. There is no guarantee that a successful species today will survive tomorrow.

The tree of life is not a divine staircase.

It is a history of survival and loss.

The Trilobites: Success Without Permanence

Trilobites were among the great success stories of ancient life.

They lived in the oceans for hundreds of millions of years. Their bodies were segmented and armored. They were diverse, widespread, and well adapted to many marine environments. They appeared in many forms and filled many ecological roles.

Then they vanished.

All of them.

The last trilobites disappeared during the Permian-Triassic extinction, the most severe mass extinction known in Earth's history.

This raises an important question for divine design.

If a creator designed trilobites so successfully, why allow every one of them to disappear? Why create such a long and diverse history only to erase the entire group? Why build a world where success for hundreds of millions of years still provides no protection from extinction?

Evolution answers easily.

Trilobites were adapted to their environments for a long time. When conditions changed beyond what their lineages could survive, they disappeared. Their long success did not guarantee permanence.

That is how natural history works.

But it does not look like a careful moral plan.

The Ammonites and the Vanished Seas

Ammonites were marine animals with coiled shells. They were abundant, diverse, and beautiful. Their fossils are found across the world, sometimes in elegant spirals that look almost designed as ornaments.

For millions of years, they lived in ancient seas.

Then they disappeared during the same event that ended the non-avian dinosaurs.

Their shells remain in stone.

The animals are gone.

A religious person may look at an ammonite fossil and admire God's artistry. But the harder question remains: why create such beauty only to erase it? Why fill ancient oceans with these creatures for vast ages, then end them completely?

The design argument often celebrates beauty when it is visible and ignores disappearance when it becomes inconvenient.

Evolution does not have that problem.

Ammonites evolved, diversified, flourished, and went extinct. Their history is not a moral story. It is a natural one.

The fossil in the hand is beautiful, but it is also a reminder.

Nature preserves no species by love.

Dinosaurs: Rulers Without Guarantee

Dinosaurs dominated many land environments for an enormous span of time.

They were not a brief experiment. They were not minor creatures on the edge of life. They were among the most successful animals in Earth's history. Some were enormous. Some were small. Some were predators. Some were plant eaters. Some had feathers. Some gave rise to birds, the surviving dinosaur lineage.

Then, about 66 million years ago, an asteroid impact and its environmental consequences helped end the age of non-avian dinosaurs.

A group that had flourished for millions of years disappeared, except for the bird lineage.

This is one of the clearest lessons in natural history:

Success does not mean safety.

A species can be powerful, widespread, and well adapted, and still vanish when the world changes suddenly.

If life were guided by perfect divine planning, why create such magnificent animals only to erase most of them long before humans appeared? Why make them dominate the Earth for immense ages if the real goal was humanity? Why use extinction as the road to later life?

Evolution answers without difficulty.

There was no guaranteed road. There was no moral script. There was changing life on a changing planet. The extinction of non-avian dinosaurs opened ecological opportunities for mammals, but that does not mean mammals were the planned purpose of dinosaur extinction.

It means history turned.

Mammoths, Saber-Toothed Cats, and Recent Loss

Extinction is not only ancient.

Some vanished animals lived close to our own time in geological terms.

Mammoths, mastodons, giant ground sloths, saber-toothed cats, and many other large animals disappeared during the late Quaternary. Climate change, human hunting, habitat change, and ecological pressure may all have played roles, depending on the species and region.

These animals are easier for us to imagine because they lived in a world closer to ours. Some early humans saw them. Some hunted them. Some painted animals on cave walls. Some lived in landscapes where huge mammals still walked.

Then many of them vanished.

Their extinction reminds us that disappearance is not only a distant fossil fact. It is a continuing possibility. Species are not protected by their size, beauty, strength, or importance to human imagination.

The mammoth did not vanish because it lacked divine purpose.

It vanished because living beings depend on conditions, and conditions change.

This is the deep logic of extinction.

No species owns the future.

Human Relatives Who Are Gone

Extinction becomes especially personal when we look at human relatives.

Homo sapiens is the only surviving human species, but we were not always alone.

There were Neanderthals. There were Denisovans. There were other members of the human family. They had bodies, tools, social lives, and histories. Some interbred with ancestors of living humans. Some may have cared for one another. Some may have buried their dead. Some may have made symbolic objects or art. They were not beasts in the crude sense. They were human relatives.

Now they are gone.

This fact deeply challenges the idea of special creation.

If humans were created as a single separate divine category, why does the fossil and genetic evidence show a branching human family? Why did other human lineages exist? Why did they disappear? Why does our own DNA carry traces of interbreeding with some of them?

The religious picture often imagines one special human origin.

The evidence shows a branching natural history.

Homo sapiens did not arrive on a clean divine stage. We emerged from a larger human family, most of which vanished.

Even our own existence is not proof of destiny.

It is the survival of one branch.

Mass Extinctions: When the World Breaks

Extinction happens in ordinary background ways, but at several points in Earth's history, it happened on a catastrophic scale.

These are mass extinctions.

During mass extinctions, large portions of life disappear in a geologically short period. Ecosystems collapse. Food chains break. Oceans change. Climate shifts. Volcanism, asteroid impact, oxygen changes, sea-level changes, and other forces may contribute.

The Permian-Triassic extinction was especially severe. It erased a vast amount of marine and terrestrial life. The Cretaceous-Paleogene extinction ended the non-avian dinosaurs and many other groups. Other mass extinctions also reshaped life dramatically.

These events are not small inconveniences for divine design.

They are enormous problems.

Why would a perfect creator build a world where life repeatedly suffers planetary-scale catastrophe? Why allow entire ecosystems to collapse? Why create complex life only to erase much of it again and again? Why make survival depend so heavily on geological accident?

Evolution has no moral difficulty here.

Life exists on a changing planet. Planets are not stable gardens. They have volcanoes, climate shifts, impacts, moving continents, changing oceans, and atmospheric changes. Life adapts when it can and dies when it cannot.

Mass extinction looks like planetary natural history.

Not divine protection.

Chance and Survival

Extinction teaches us another humbling fact:

Survival is often shaped by chance.

This does not mean everything is random. Adaptation matters. Traits matter. Environment matters. But chance also matters deeply.

An asteroid impact can change the future of life. A volcanic event can alter climate. A small population can survive in one region while another disappears. A species may live in the right place at the right time. Another may be trapped in the wrong place. Some lineages survive catastrophe not because they are morally better or perfectly designed, but because their traits and circumstances happen to fit the new world.

This is hard for religious thinking because religion often wants meaning in survival.

The survivor is blessed.

The extinct is part of a plan.

The preserved is favored.

The destroyed had a purpose we cannot see.

But science sees no such moral pattern.

Survival is not holiness.

Extinction is not sin.

The fossil record does not show divine reward and punishment. It shows biology meeting changing conditions under the influence of contingency.

Chance is not God.

Chance is part of nature.

No Moral Meaning in Extinction

Religious people sometimes interpret events morally.

A disaster becomes punishment. A recovery becomes blessing. Survival becomes evidence of favor. Death becomes part of a hidden plan.

But extinction has no moral pattern.

The trilobites did not sin. The ammonites did not disobey God. The dinosaurs were not spiritually corrupt. The mammoths were not punished. Neanderthals were not less worthy of existence.

They disappeared because the natural conditions that allowed them to continue changed or failed.

This is important because it removes a cruel burden from life.

When suffering is called punishment, the victim is morally stained. When extinction is seen as divine judgment, nature becomes a courtroom. But there is no evidence for such judgment in the fossil record.

Extinction is not moral.

It is biological.

This is a more honest view and, in many ways, a kinder one. It does not blame the suffering creature. It does not invent guilt where there is only vulnerability.

The vanished species were not cursed.

They were natural beings in a changing world.

Why Extinction Is a Problem for Perfect Design

Extinction is difficult for divine design for several reasons.

First, it shows waste on an enormous scale. Countless species appear, survive for a time, and disappear. If God created each species with purpose, why erase most of them?

Second, extinction shows that adaptation is temporary. A species may be well suited to one environment and helpless in another. This fits evolution, not perfect planning.

Third, extinction shows that life is not human-centered. Most extinction occurred before humans existed. Vast histories of life began and ended with no human witness.

Fourth, extinction shows that nature does not preserve beauty, complexity, or innocence. Beautiful animals vanish. Complex ecosystems collapse. Young die with old. There is no moral filter.

Fifth, mass extinctions show that the planet itself is not arranged as a safe home for life. It is dynamic, unstable, and sometimes catastrophic.

A perfect designer could have made a world without mass extinction.

Evolution cannot promise that.

The fossil record therefore gives us not only evidence for evolution, but evidence against perfect design.

The Flood Explanation Fails

Some religious traditions try to explain fossils and extinction through a great flood.

This does not work scientifically.

The fossil record is not a chaotic mixture of all creatures buried at once. It shows order through geological layers. Different fossils appear in different periods. Marine fossils, land animals, plants, early vertebrates, dinosaurs, mammals, and humans do not appear in one simple flood deposit. The sequence fits deep time and evolution, not a single recent catastrophe.

Geology also shows many kinds of deposits formed in many environments over vast time: deserts, rivers, lakes, reefs, volcanic ash, deep oceans, shallow seas, floodplains, glaciers, and more. A single flood cannot explain the complexity of Earth's rock record.

The flood explanation is an attempt to force evidence into scripture.

It does not arise naturally from the evidence.

Science does not need to protect a sacred story. It can allow the rocks to speak. And the rocks speak of deep time, changing environments, extinction, and evolution.

Not a recent global flood.

The Human-Centered Story Breaks Down

Extinction also weakens the idea that life's history was arranged for human beings.

For billions of years, no humans existed. For hundreds of millions of years, life flourished and vanished without us. Entire worlds came and went before the first human ancestor walked upright. Dinosaurs lived for far longer than humans have existed. Trilobites lasted far longer than human civilization. Countless species lived complete histories with no connection to human religion.

If humans were the central purpose of creation, the method seems strange.

Why such a long pre-human history?

Why so much suffering before humans?

Why so many vanished species?

Why so many mass extinctions?

Why such a late arrival for the supposed purpose of it all?

Evolution answers simply.

Humans are late because we evolved late.

We are not the central purpose of life's history. We are one recent branch. Our existence is remarkable, but it is not evidence that the entire history of life was aimed at us.

Human pride wants the universe to be about us.

Science teaches humility.

Extinction and the Fragility of Life

Extinction shows that life is fragile.

A species may seem secure, but conditions can change. A habitat can disappear. A climate can shift. A disease can spread. A predator or competitor can arrive. A food source can vanish. Human activity can destroy ecosystems. Once the last individual of a species dies, an ancient line of life ends forever.

This is not just a fact about the past.

It is a warning for the present.

Human beings now have the power to drive extinction through habitat destruction, climate change, pollution, hunting, invasive species, and ecological disruption. If there is no divine plan protecting the living world, then responsibility falls on us.

Science-based atheism does not lead to indifference.

It leads to responsibility.

The living world is not safe because God guards it. Species are not protected by sacred purpose. Nature does not guarantee recovery.

If we value life, we must protect it.

There is no evidence that heaven will do the work for us.

Extinction Does Not Destroy Meaning

Some readers may feel that extinction makes life seem hopeless.

If species disappear, if worlds vanish, if nature does not preserve beauty or innocence, what meaning can life have?

The answer is that meaning does not require permanence.

A human life is meaningful though it is temporary. A flower is beautiful though it fades. A song matters though it ends. A friendship matters though life is brief. A species matters though it may not last forever.

Meaning is not the same as eternal survival.

In fact, impermanence can make life more precious.

The fossil record teaches that life is not guaranteed. It tells us that every living form is part of a fragile chain. It tells us that survival is not a divine right. It tells us that the present world is precious because so much has been lost.

Religion often tries to comfort us with permanence.

Science gives us something different:

Clarity.

Life is temporary.

That does not make it worthless.

It makes it urgent.

The Design Argument Cannot Count Only Survivors

The design argument often looks at living species and says, “See how well designed they are.”

But this is a biased view.

It counts the survivors and ignores the dead.

It is like judging a war by interviewing only those who came home. It is like judging a failed business world by looking only at companies that survived. It is like praising a lottery system because a few people won, while ignoring millions who lost.

The living world looks more designed when we forget extinction.

Once extinction is included, the picture changes.

Life is not a smooth collection of successful designs. It is a long history of experiments, dead ends, catastrophes, and survival. The species alive today are not proof that the whole system was perfectly planned. They are the few branches that reached the present.

Evolution explains this naturally.

Design thinking hides the graveyard.

Why "God Had a Reason" Is Not Enough

When confronted with extinction, some believers say, "God had a reason."

Perhaps. But that is not science.

A statement that explains everything explains nothing. If survival is God's reason and extinction is also God's reason, then the claim cannot be tested. If every fossil fits God's plan, no matter what it shows, then the plan is not a scientific explanation.

Science asks better questions.

What caused the extinction?

What environmental changes occurred?

Which species survived?

Which traits helped survival?

What does the timing show?

What do fossils reveal?

How did ecosystems recover?

What new lineages appeared afterward?

These questions lead to knowledge.

"God had a reason" leads to silence.

This is why divine design fails as science. It does not help us understand the fossil record. It does not explain the order of fossils. It does not explain mass extinction. It does not predict what we should find in ancient rocks.

Evolutionary science does.

The Graveyard Speaks

The fossil record is not merely a collection of dead bones.

It is testimony.

It says that life is old.

It says that life has changed.

It says that species are temporary.

It says that extinction is common.

It says that humans arrived late.

It says that nature has no permanent favorites.

It says that survival depends on conditions, not divine love.

This testimony is powerful because it comes not from philosophy, but from the Earth itself.

The rocks do not preach atheism. They do not argue. They do not mock religion. They simply preserve the evidence.

And the evidence does not support special creation.

It supports natural history.

If God designed life perfectly, the fossil record is strange, wasteful, and morally difficult. If life evolved on a changing planet, the fossil record makes sense.

The graveyard speaks more clearly than the sermon.

Extinction and Science-Based Atheism

Extinction is central to science-based atheism because it removes the illusion of divine purpose from life's history.

The old story says life was created with intention.

Extinction shows vast loss without moral pattern.

The old story says humans are the center.

Extinction shows billions of years of life before humans.

The old story says nature reflects divine wisdom.

Extinction shows dead branches, failed lineages, mass death, and chance survival.

The old story says God designed living beings.

Extinction asks why most designs were erased.

A believer may still hold faith privately. But as a scientific explanation, divine design becomes weaker with every fossil layer.

The history of life does not need God to make sense.

It needs time, change, inheritance, environment, catastrophe, adaptation, and extinction.

That is enough.

The Calm Conclusion

Extinction is one of the strongest arguments against perfect divine design.

It shows that life is not a fixed creation. It shows that species are temporary. It shows that nature does not preserve beauty, complexity, innocence, or strength. It shows that the history of life is filled with disappearance.

It shows that humans are late arrivals in an ancient, unstable world.

The fossil record is not a record of perfect planning.

It is a record of life trying, changing, surviving, and dying.

Some branches continued.

Most ended.

This is exactly what evolution predicts.

It is not what perfect divine design would lead us to expect.

A perfect designer could have created life without such waste. A loving creator could have built a world without mass extinction. An all-knowing planner could have avoided billions of years of vanished life before humans arrived.

But the Earth does not show that kind of plan.

It shows natural history.

The conclusion is unavoidable:

Extinction is not evidence of divine design.

It is evidence that life is fragile, temporary, and evolved.

No designer is required.

Chapter Summary

Extinction is not rare in the history of life. Most species that ever lived are gone.

The tree of life contains many dead branches, not a smooth ladder of progress.

Trilobites, ammonites, non-avian dinosaurs, mammoths, saber-toothed cats, and many human relatives are extinct.

Mass extinctions repeatedly reshaped life on Earth.

Survival is not moral, and extinction is not divine punishment.

Extinction existed long before humans and cannot be explained scientifically as a result of human sin.

The fossil record does not support a single recent flood. It supports deep time, sequence, extinction, and evolution.

Extinction weakens the claim that life's history is human-centered or divinely planned.

The design argument often counts only surviving species and ignores the vast graveyard of vanished life.

Evolution explains extinction far better than perfect divine design.

Key Takeaway

The history of life is not a smooth divine plan. It is filled with vanished species, dead branches, mass extinctions, and chance survival. Most life that ever existed is gone.

This pattern makes sense in evolution, where species survive only as long as conditions allow. It does not look like the work of a perfect, loving designer. The fossil record is a graveyard of divine design.

Sources

Charles Darwin. On the Origin of Species.

National Academies of Sciences, Engineering, and Medicine. Science, Evolution, and Creationism.

University of California Museum of Paleontology. Understanding Evolution: Extinction.

University of California Museum of Paleontology. Understanding Evolution: Mass Extinction.

Natural History Museum. What Killed the Dinosaurs?

United States Geological Survey. Mesozoic Era and Cretaceous Period Resources.

Smithsonian National Museum of Natural History. Human Origins Program: Neanderthals.

Smithsonian National Museum of Natural History. Human Origins Program: Early Human Species.

Donald R. Prothero. Evolution: What the Fossils Say and Why It Matters.

Stephen Jay Gould. Wonderful Life.

Peter Ward. Under a Green Sky.

Douglas H. Erwin. Extinction: How Life on Earth Nearly Ended 250 Million Years Ago.

Carl Zimmer. Evolution: The Triumph of an Idea.

Transition to the Next Chapter

Extinction shows that life is not a perfect plan.

But one question remains especially emotional for religious readers:

What about us?

Were human beings specially created, separate from all other life? Or are we also part of the same natural story?

The next chapter will examine human evolution, not as an insult to human dignity, but as one of the strongest truths science has revealed about who we are.

Chapter 15: Human Evolution: No Special Creation, No Loss of Dignity

The Mirror That Frightened Religion

Few scientific truths have disturbed religion more than this one:

Human beings are animals.

Not ordinary animals. Not simple animals. Not animals without language, memory, love, art, grief, morality, science, imagination, and self-awareness.

But animals nonetheless.

We are mammals.

We are primates.

We are great apes.

We are part of the tree of life.

This truth has frightened many people because it touches human pride. Religion often tells us that we were specially created by God, placed above the rest of nature, made in the divine image, and given a soul that separates us from every other living being.

Evolution tells a different story.

It tells us that our bodies were not made separately from the bodies of other animals. Our hands, eyes, teeth, spine, blood, brain, genes, and bones carry the marks of ancestry. Our history is not a sudden divine act. It is a long natural process.

To some people, this feels like a loss.

But it does not have to be.

Human evolution does not make us worthless. It makes us honest. It places us inside the living world, not above it by supernatural decree. It gives us kinship instead of isolation. It replaces sacred pride with scientific humility.

We do not need special creation to have dignity.

We need consciousness, compassion, reason, vulnerability, memory, and the ability to care for one another.

Those are enough.

We Did Not Evolve From Modern Chimpanzees

One misunderstanding must be corrected at once.

Humans did not evolve from modern chimpanzees.

This false claim is often repeated by people who misunderstand evolution or wish to ridicule it. They ask, “If humans came from monkeys, why are monkeys still here?”

But that is not what evolution says.

Humans and modern chimpanzees share common ancestors. We are not descended from chimpanzees living today. Rather, humans and chimpanzees are like cousins whose family lines split from an earlier ancestral population.

The same is true in ordinary families. You are not descended from your cousin. You and your cousin share grandparents or earlier ancestors. Your cousin can exist at the same time you do because both of you come from a shared family history.

Evolution works in a similar way.

After the ancestral lines split, one lineage led toward modern chimpanzees and bonobos, while another lineage led through many forms toward modern humans. Neither modern species is the ancestor of the other.

This point matters because false versions of evolution make it easier for religious critics to reject it. A serious argument must describe the science correctly.

Evolution does not say modern apes turned into humans.

It says humans are one branch of the ape family.

The Human Body Remembers Its Ancestry

The human body is full of signs that we belong to nature.

We have a backbone, like other vertebrates.

We have hair, mammary glands, and three middle ear bones, like other mammals.

We have forward-facing eyes, grasping hands, nails instead of claws, and flexible shoulders, like other primates.

We share many anatomical and genetic features with other great apes.

These are not superficial coincidences.

They are patterns of ancestry.

If humans were separately created from nothing, why do our bodies fit so clearly within the mammalian and primate pattern? Why do we share skeletal structures with other animals? Why do our embryos develop in ways that reveal deep biological kinship? Why does our DNA place us among the great apes? Why do our bodies contain inherited weaknesses similar to those of other animals?

Separate creation does not explain these facts well.

It only says, "God chose to make us that way."

Evolution gives a better answer.

We resemble other animals because we are related to them.

Our bodies are not separate divine inventions. They are modified versions of older biological forms. We are not strangers in nature. We are descendants.

That does not make us less human.

It makes our humanity part of a larger story.

Walking Upright

One of the most important changes in human evolution was walking upright.

Bipedalism changed everything. It freed the hands. It changed the pelvis, spine, legs, feet, and balance. It allowed carrying food, tools, and children. It may have helped our ancestors move efficiently across certain landscapes. It changed how we lived in the world.

But walking upright was not a perfect gift.

It came with costs.

The human spine became vulnerable to back pain. The pelvis had to balance the demands of upright walking with childbirth. Knees, hips, ankles, and feet carried new stresses. A body shaped from earlier primate anatomy had to support a new way of moving.

This is exactly what evolution predicts.

A designer starting from scratch could build an upright creature with fewer mechanical problems. Evolution could not start from scratch. It modified an inherited body plan.

That is why human walking is both remarkable and flawed.

We move on two legs, but we carry the price of that history in our joints, spine, and childbirth.

The body tells the story.

Hands, Tools, and Intelligence

Human hands are extraordinary.

They can hold a scalpel, write a sentence, build a shelter, paint a picture, play a musical instrument, comfort a child, prepare food, repair a machine, and perform surgery. The human hand is one of the great instruments of culture.

But the hand did not descend from heaven.

It is a modified primate hand.

Our ancestors inherited grasping hands from earlier primates. Over time, changes in anatomy, posture, brain control, and behavior allowed the hand to become a powerful tool-making and tool-using structure.

Tool use did not require special creation. Other animals use tools too, though not with the same complexity as humans. Chimpanzees, crows, otters, and some other animals use objects in useful ways. Human beings took this capacity to an extraordinary level, especially through culture, language, memory, and teaching.

This is an important point.

Human uniqueness did not appear from nothing. It grew from older capacities.

Grasping became manipulation.

Manipulation became tool use.

Tool use became culture.

Culture changed survival.

Learning became inheritance of a different kind.

Evolution shaped the body. Culture then shaped the human world.

No divine creation is needed to explain this.

The Growing Brain

The human brain is astonishing.

It allows language, memory, planning, imagination, moral reflection, mathematics, music, grief, humor, and science. It allows a universe of thought inside a skull.

For religion, the brain has often seemed like evidence of the soul. How else, people ask, could matter think? How could neurons produce consciousness? How could a biological organ produce poetry, love, and philosophy?

Book 5 of this series will examine mind and consciousness more deeply. Here, the important point is simpler:

The human brain has an evolutionary history.

Brain size and organization changed across the human lineage. Not in a straight, simple ladder, but in a branching and complex pattern. Different human relatives had different brain sizes, bodies, tools, and ways of living. Modern humans inherited a brain shaped by survival, social life, communication, tool use, memory, cooperation, and culture.

The brain did not arrive fully formed as a supernatural object.

It developed through natural history.

This does not reduce thought to meaninglessness. It simply says that thought depends on biology. Damage the brain, and memory can disappear. Change brain chemistry, and mood can change. Injure certain areas, and speech can fail. Disease the brain, and personality can alter.

The evidence points strongly to mind as a function of the brain, not as a ghost placed inside the body.

Human intelligence is extraordinary.

But extraordinary does not mean supernatural.

Fossils of the Human Family

The fossil record of human evolution does not show one sudden leap from animal to modern human.

It shows a branching family.

There were australopithecines, early members of the genus Homo, Homo erectus, Neanderthals, Denisovans, and other human relatives. Some walked upright. Some used tools. Some had larger brains than earlier ancestors. Some spread across great distances. Some lived in harsh climates. Some likely cared for one another. Some may have made symbolic objects, used pigments, or buried their dead.

The details are complex, and new discoveries continue to refine the story. That is how science works.

But the broad message is clear:

Humans have a natural history.

We were not created suddenly in our present form.

Our ancestors changed over time. Some branches ended. Some interbred. Some left genetic traces in living people. The human family tree is not a simple straight line, and it is certainly not a sacred story of one specially created pair.

This deeply weakens religious claims of special creation.

If humans were created separately and uniquely, why does the evidence show a branching human family? Why do we find fossil forms with mixtures of human-like and ape-like features? Why do we find ancient human relatives who lived, adapted, and vanished? Why does DNA show interbreeding with Neanderthals and Denisovans?

Evolution explains these facts.

Special creation does not.

Neanderthals and the Problem of Human Uniqueness

Neanderthals are especially important because they disturb simple religious categories.

They were not modern humans, but they were not mere beasts. They had large brains. They made tools. They survived in difficult environments. They likely cared for injured members of their groups. Evidence suggests symbolic behavior in at least some settings. They interbred with ancestors of many living humans, leaving genetic traces in people today.

So where would religion place them?

Did they have souls?

Were they made in God's image?

Were they moral beings?

Were they animals only?

Were they human enough for heaven or sin?

These questions show how poorly simple religious categories fit the evidence.

Evolution has no difficulty here.

Neanderthals were a human lineage, closely related to us, with their own history. They were part of the branching story of human evolution. They do not need to be forced into a theological box.

Their existence shows that humanity did not appear as a sharp supernatural category.

It emerged gradually, with relatives, branches, overlaps, and extinctions.

The boundary between "human" and "not human" is not as clean as religion often imagines.

Nature is historical.

Denisovans and the Hidden Human Past

Denisovans remind us that the human story is still being uncovered.

For a long time, they were unknown. Then genetic evidence from ancient remains revealed another human lineage, related to Neanderthals and modern humans. Denisovan ancestry survives in some living populations.

This is a remarkable scientific achievement.

From small fragments of bone and DNA, scientists uncovered a hidden branch of the human family.

What does this mean for the religious story?

It means human origins are far more complex than the idea of one specially created pair. Our past includes multiple human lineages, migration, interbreeding, extinction, and survival. The story is not clean, simple, or scriptural.

It is biological.

Science does not know every detail yet. New fossils and genetic discoveries will continue to refine the story. But uncertainty in details does not rescue special creation. The broad evidence still points toward natural human evolution.

Denisovans show that the human family tree is larger than we once knew.

That is not a problem for evolution.

It is exactly the kind of discovery evolution can absorb and explain.

There Was No First Perfect Human Pair

Many religious traditions depend on the idea of an original human pair.

A first man.

A first woman.

A special creation.

A fall.

A moral drama beginning with two divinely made humans.

But evolution does not work that way.

Species arise from populations, not from a single sudden pair created from nothing. There may be individuals who are ancestral to many later people, but that is not the same as a magically created first couple. The transition from earlier human ancestors to modern humans was gradual and population-based.

There was no clear moment when two non-humans gave birth to a fully modern human in the religious sense.

There was no single morning when biology stopped being animal and became sacred humanity.

Instead, there were populations changing over time. Anatomy changed. Brains changed. Behavior changed. Tools changed. Language likely developed gradually. Culture became more complex. Symbolic thought deepened. Social life expanded.

Humanity emerged.

It was not inserted.

This is a major challenge to religious doctrines that depend on special creation.

If there was no first perfect pair, many related theological claims become much harder to defend scientifically.

Human Chromosome 2 and the Mark of Kinship

DNA has made human evolution even harder to deny.

Humans have 46 chromosomes. Chimpanzees, bonobos, gorillas, and orangutans have 48. If humans and other great apes share ancestry, then science would expect an explanation for this difference. One possibility is that two ancestral chromosomes fused in the human lineage.

That is what the evidence shows.

Human chromosome 2 contains signs consistent with an ancient fusion event. It carries telomere-like sequences in an internal position, where chromosome ends appear to have joined. It also contains evidence of a second centromere region that is no longer active in the usual way.

This is not a poetic argument.

It is a biological clue.

If humans were separately created, why would one of our chromosomes look like two ancestral ape chromosomes fused together? Why would the evidence match the prediction of common ancestry?

Separate creation has no strong answer.

Evolution has a clear one.

Our genome carries our history.

It shows that human beings are not outside the ape family.

We are part of it.

The Image of God and the Evidence of Biology

Religion often claims that humans are made in the image of God.

This phrase has deep emotional power. It gives people dignity, purpose, and a feeling of sacred identity. It tells them they are not merely part of nature, but reflections of the divine.

But as a scientific claim, it has no evidence.

What would it mean biologically to be made in the image of God? Does it refer to anatomy? If so, why do our bodies resemble other primates? Does it refer to intelligence? If so, why do intelligence, memory, emotion, and problem-solving appear in degrees across animals? Does it refer to morality? If so, why do moral instincts have roots in social life, empathy, cooperation, attachment, and group survival?

The phrase may have poetic meaning, but it does not explain human origins.

Biology does.

Human beings are not shaped like gods. We are shaped like evolved primates. We have mammalian organs, primate hands, ape ancestry, vulnerable bodies, and brains dependent on blood, oxygen, glucose, sleep, chemistry, and development.

The evidence does not point to divine image.

It points to evolutionary history.

Morality Without Special Creation

Some people fear that if humans evolved, morality collapses.

They think that if we are animals, then anything is allowed. They think morality requires a soul or divine command.

This fear is mistaken.

Human morality does not require special creation. It grows from our nature as social, conscious beings. We can suffer. We can recognize suffering in others. We form bonds. We raise children. We cooperate. We remember harm. We respond to fairness and unfairness. We build rules because social life requires trust.

Evolution may help explain the roots of empathy, cooperation, kin care, reciprocity, guilt, shame, and group loyalty. Culture, reason, education, law, and moral reflection can then expand and refine these instincts.

This does not mean nature itself is moral. As an earlier chapter showed, nature contains much cruelty. But human beings can rise above blind nature precisely because we can think, feel, and choose.

Atheism does not destroy morality.

It makes morality our responsibility.

If no god commands kindness from the sky, then we must build kindness here on Earth.

That is not moral emptiness.

It is moral maturity.

Dignity Without a Soul

Another fear is that evolution removes the soul and therefore removes dignity.

But dignity does not need a supernatural soul.

A person with Alzheimer's disease still has dignity, even if memory fades because the brain is diseased. A newborn has dignity before having fully developed reasoning. A person under anesthesia has dignity though consciousness is temporarily absent. A severely disabled person has dignity even if many mental functions are impaired.

Dignity does not depend on proving an invisible soul.

It depends on recognizing that conscious and vulnerable beings can be harmed, loved, protected, neglected, respected, or abused.

Human dignity comes from our capacity for experience, suffering, relationship, memory, hope, and care. It comes from the fact that each life is lived from the inside. Each person has a point of view, a body that can feel pain, and relationships that matter.

Evolution does not take this away.

It explains how beings capable of dignity came to exist.

A natural origin does not remove value.

A child born through biology is not less precious than a child imagined as magically created.

Human beings are precious because life is precious, consciousness is precious, and suffering matters.

The Emotional Resistance to Human Evolution

Human evolution faces more resistance than the evolution of plants, insects, or bacteria because it touches identity.

Many people can accept that bacteria evolve. They may accept that dogs and wolves are related. They may accept that birds descended from earlier forms. But when the same logic reaches humans, they become uncomfortable.

The reason is clear.

Human evolution removes us from the religious throne.

It says we are not separate creations. It says we did not arrive fully formed from divine intention. It says our bodies are related to other animals. It says our minds depend on evolved brains. It says our moral capacities have natural roots.

This can feel like humiliation if one has been taught that dignity requires divine separation.

But it can feel like liberation if one values truth.

We do not need to pretend to be supernatural in order to matter. We do not need to deny our ancestry to preserve our worth. We do not need to stand outside nature to live meaningful lives.

We can accept that we are evolved and still love, create, heal, teach, protect, and search for truth.

Indeed, accepting truth is part of human dignity.

The Human Animal Is Still Extraordinary

Saying humans are animals does not mean humans are unremarkable.

We are extraordinary in natural ways.

We use complex language. We pass knowledge across generations. We create moral systems. We imagine the future. We perform surgery. We write music. We study galaxies. We bury our dead. We care for the sick. We build schools, hospitals, libraries, laws, and scientific institutions. We ask why we exist.

These are remarkable achievements.

But none requires special creation.

They arise from evolved brains, social cooperation, symbolic language, culture, memory, and cumulative learning. Evolution gave us the biological foundation. Culture built upon it.

This is a powerful story.

The religious story says God made humans special.

The scientific story says humans became special through nature, time, brain, language, cooperation, and culture.

The scientific story is not smaller.

It is richer because it shows the path.

It explains how a primate lineage became capable of understanding its own origin.

The Medical Evidence of Our Animal Nature

Medicine also reveals our animal nature.

Human anatomy is mammalian anatomy. Our organs can be studied in relation to other animals. Medical research often uses animal models because biological systems are shared. Drugs, hormones, immune pathways, genes, infections, and organ functions have deep evolutionary connections.

This would make little sense if humans were biologically separate creations.

It makes perfect sense if we share ancestry with other life.

The human heart is not a supernatural organ. It is a mammalian pump. The lungs, kidneys, liver, muscles, bones, nerves, and blood belong to the natural world. They obey chemistry and physiology. They fail in natural ways. They heal in natural ways. They respond to natural treatment.

A physician does not treat the body as a sacred exception to biology.

A physician treats it as biology.

That does not make care less compassionate.

It makes care possible.

Medicine works because humans are natural beings.

Human Evolution and the Death of Special Creation

Human evolution is a direct challenge to special creation.

Special creation says humans were made separately by God.

Evolution says humans emerged from earlier life through natural processes.

Special creation says humans stand apart from animals.

Evolution says humans are animals with distinctive traits.

Special creation says our bodies were divinely formed.

Evolution says our bodies carry inherited anatomy.

Special creation says our origin is supernatural.

Evolution says our origin is natural.

These two views cannot both be scientifically correct.

One is supported by fossils, anatomy, genetics, embryology, comparative biology, and medicine.

The other is supported by scripture, tradition, and faith.

For a science-based reader, the conclusion should be clear.

Human beings were not specially created by God.

We evolved.

This does not disprove every possible idea of God in a philosophical sense. But it does remove God as a scientific explanation for human origins. There is no scientific evidence that humans were specially created. There is strong scientific evidence that humans are part of the evolutionary history of life.

That is enough for the purpose of this book.

No Special Creation, No Despair

Some people think that if humans are not specially created, life becomes empty.

But why should that be true?

Is a child less precious because he grew in a womb through biology? Is love less real because the brain and body are involved? Is music less moving because sound waves obey physics? Is a sunset less beautiful because it is caused by light scattering in the atmosphere?

Natural explanation does not destroy meaning.

It removes illusion.

Human beings do not need to be molded by divine hands to matter. We matter because we are conscious, vulnerable, social, thinking beings. We matter because we can experience joy and suffering. We matter because we can harm or help one another. We matter because our lives are brief and irreplaceable.

Evolution gives us a natural origin.

It does not take away our human value.

In fact, it may make our value more urgent. If there is no divine plan protecting us, then our care for one another becomes more important. If there is no afterlife guaranteed by a creator, then this life becomes more precious. If there is no heavenly justice, then justice on Earth matters more.

Atheism is not despair.

It is responsibility.

We Are Life Becoming Aware of Itself

There is a beautiful way to understand human beings without God.

We are life becoming aware of itself.

For billions of years, life changed without knowing its own history. Cells divided. Organisms reproduced. Species appeared and vanished. Forests grew. Seas filled with life. Animals hunted, fled, suffered, and survived. No one wrote it down. No one understood the process.

Then, very late in the story, one species evolved a brain capable of asking questions.

Where did we come from?

Why do we suffer?

What is life?

What is death?

Are we alone?

Is there a God?

How do we know what is true?

Religion gave early answers, but science gave better ones.

Human beings are the part of nature that can study nature. We are not above the tree of life. We are a branch that learned to look back at the tree.

That is profound.

It does not require a soul. It does not require special creation. It does not require divine breath.

It requires evolution.

The Calm Conclusion

Human evolution is not an insult to humanity.

It is the truth of our origin.

We did not evolve from modern chimpanzees. We share common ancestors with them. We are primates, mammals, vertebrates, animals, and living beings connected to the rest of life through deep ancestry. Our bodies carry evolutionary history. Our fossils show a branching human family. Our DNA places us among the great apes. Our brain, language, tools, culture, and morality developed through natural processes.

This truth removes the need for special creation.

But it does not remove dignity.

We do not need to be made in the image of God to be valuable. We do not need supernatural souls to deserve compassion. We do not need divine origin to love, think, suffer, heal, create, and care.

Human beings are special in natural ways.

We are evolved life capable of understanding evolution.

That is enough.

The religious story says we were created separately from nature.

Science says we grew out of nature.

The scientific story is not a fall from dignity.

It is a rise into honesty.

No special creation is required.

No divine image has been demonstrated.

No God is needed to explain who we are.

Chapter Summary

Humans did not evolve from modern chimpanzees. Humans and chimpanzees share common ancestors.

Human beings are mammals, primates, and great apes.

The human body carries clear signs of evolutionary ancestry.

Walking upright gave advantages but also created costs, including spinal, pelvic, and childbirth problems.

Human hands, tools, brain growth, language, and culture developed through natural history.

The fossil record shows a branching human family, not sudden special creation.

Neanderthals and Denisovans show that human history is more complex than a simple religious origin story.

Human chromosome 2 carries evidence consistent with an ancient fusion event.

Evolution does not support the idea of one specially created first human pair.

Human dignity does not require a soul or divine image.

Human morality can be understood through social life, empathy, reason, culture, and cooperation.

Human evolution removes God as a scientific explanation for human origins.

Key Takeaway

Human beings were not specially created outside nature. We are evolved animals, part of the primate family and the larger tree of life. This truth does not destroy dignity. It replaces supernatural pride with natural kinship, humility, and responsibility. We do not need divine creation to matter. We matter because we are conscious, vulnerable, social beings capable of love, thought, suffering, and care.

Sources

Charles Darwin. The Descent of Man.

Smithsonian National Museum of Natural History. Human Origins Program.

Smithsonian National Museum of Natural History. Ancient DNA and Neanderthals.

National Academies of Sciences, Engineering, and Medicine. Science, Evolution, and Creationism.

National Human Genome Research Institute. Human Chromosome 2 Resources.

University of California Museum of Paleontology. Understanding Evolution: Human Evolution.

Richard Leakey. The Origin of Humankind.

Donald Johanson and Maitland Edey. Lucy: The Beginnings of Humankind.

Chris Stringer. Lone Survivors: How We Came to Be the Only Humans on Earth.

Svante Pääbo. Neanderthal Man: In Search of Lost Genomes.

Transition to the Next Chapter

Human evolution removes one of religion's most cherished claims: that we were specially created by God.

But religious belief rarely disappears after one explanation fails. It often retreats into the remaining gaps.

If evolution is true, some will say God guided it. If DNA shows common ancestry, some will say God wrote the code. If disease and extinction weaken design, some will say God's plan is mysterious.

The next chapter will examine these last hiding places of the designer.

Chapter 16: The Last Hiding Places of the Designer

The God Who Keeps Moving

There is a familiar pattern in the history of religion.

When thunder was not understood, gods were placed in the storm.

When disease was not understood, spirits, curses, or divine punishment were placed in the body.

When the planets were not understood, divine hands were placed in the heavens.

When species were not understood, God was placed in the origin of animals and plants.

When consciousness was not understood, God was placed inside the soul.

Again and again, the unknown became a home for God.

Then science advanced.

Lightning became electricity.

Disease became microbiology, genetics, pathology, and physiology.

Planetary motion became astronomy, gravity, and physics.

Species became evolution and common ancestry.

Human beings became biology, brain, culture, and history.

Each time science explained more, God moved.

Not necessarily out of every human heart, but out of the explanation.

God moved from thunder to mystery.

From disease to divine plan.

From special creation to guided evolution.

From the body to the soul.

From direct action to invisible influence.

From evidence to faith.

This chapter is about those last hiding places.

By this point in the book, the biological case is strong. Life looks designed because natural selection can produce adaptation. Fossils show deep time and change. DNA reveals common ancestry. The tree of life connects living beings into one family. The body shows tinkering, not perfect engineering. Disease challenges loving design. Nature shows suffering and indifference. Extinction shows no smooth divine plan. Human evolution removes special creation.

Yet belief often remains.

It retreats into objections.

"Evolution is only a theory."

"Small evolution happens, but large evolution does not."

"There are no transitional fossils."

"Random chance cannot create life."

"The eye could not have evolved."

"DNA proves a divine programmer."

"Irreducible complexity proves design."

"God used evolution."

"Science cannot disprove God completely."

These arguments may sound strong to believers, especially to those who have heard them many times in religious settings. But most of them are not strong scientifically. They survive because they protect an emotional need.

The designer keeps moving because the evidence keeps pushing him back.

"Evolution Is Only a Theory"

One of the most common objections is also one of the weakest:

"Evolution is only a theory."

This statement depends on a misunderstanding of the word theory.

In everyday conversation, a theory may mean a guess. Someone may say, "I have a theory about why he was late," meaning a suspicion or personal idea.

But in science, a theory is not a casual guess. A scientific theory is a broad explanation supported by evidence, tested against observations, and useful for making sense of many facts.

Germ theory explains infectious disease.

Cell theory explains the basic unit of living organisms.

Atomic theory explains matter.

Gravitational theory explains attraction between masses.

Evolutionary theory explains the diversity, adaptation, and relatedness of life.

Calling evolution a theory does not weaken it.

It places it among the major explanatory frameworks of science.

Evolution is supported by fossils, DNA, anatomy, embryology, biogeography, observed evolution, comparative genomics, and natural selection. It is not an atheist slogan. It is not a religious belief. It is not a nineteenth-century guess frozen in time. It is the foundation of modern biology.

A person may reject evolution for religious reasons, but calling it "only a theory" does not make that rejection scientific.

The real question is not whether evolution is called a theory.

The real question is whether evolution explains the evidence.

It does.

Divine design does not explain it better.

"Microevolution, But Not Macroevolution"

Another common retreat says:

Small changes happen, but large evolution does not.

This is often stated as, "I accept microevolution, but not macroevolution."

At first, this sounds reasonable. Many people can accept that bacteria become resistant to antibiotics, insects become resistant to pesticides, or finch beaks change under environmental pressure. But they resist the idea that such changes can accumulate into larger transformations over long periods.

The problem is that this creates an artificial wall.

Microevolution refers to smaller changes within populations, such as changes in gene frequencies or traits across generations. Macroevolution refers to larger patterns over longer time, including speciation, major differences among groups, and large-scale evolutionary history.

But macroevolution is not a separate miracle. It is not a completely different process. It is evolutionary change seen across longer time, larger populations, branching lineages, isolation, adaptation, and extinction.

A stream can cut a small groove after a storm. Given enough time, water can carve a canyon. The small process and the large result are connected.

Similarly, small evolutionary changes can accumulate. Populations can become separated. They can adapt to different environments. Genetic differences can grow.

Reproductive barriers can develop. Lineages can split. Over deep time, major differences can appear.

A person who accepts variation, inheritance, mutation, selection, and population change has already accepted the machinery of evolution.

Rejecting the larger result while accepting the mechanism is often not a scientific position.

It is a religious boundary drawn where the evidence becomes emotionally uncomfortable.

"There Are No Transitional Fossils"

The claim that there are no transitional fossils has been repeated so often that many people believe it without examination.

But it is false.

There are many transitional fossils.

A transitional fossil is not a crude cartoon creature, half one modern animal and half another. It is a fossil that shows a mixture of traits connecting earlier and later groups, or helps document a major evolutionary transition.

Archaeopteryx shows a mixture of bird-like and dinosaur-like traits.

Tiktaalik helps illuminate the transition from fish-like ancestors toward early land vertebrates.

Early whale fossils show stages in the transition from land mammals to aquatic whales.

Feathered dinosaurs show connections between dinosaurs and birds.

Human fossils show a branching history of human relatives, not sudden special creation.

The fossil record is not complete. No honest scientist claims it is. Fossilization is rare. Many organisms decay without preservation. Many fossils are destroyed. Many remain undiscovered. Soft-bodied organisms preserve poorly. Geological processes erase evidence.

But incompleteness is not absence.

A family album may be missing many photographs, but if it shows grandparents, parents, children, and cousins in meaningful sequence, it still provides evidence of family history.

The fossil record is like that.

It does not contain every life form, but it contains enough to show pattern, sequence, transition, extinction, and change.

The claim that there are no transitional fossils is not a serious scientific objection.

It is a slogan.

"Random Chance Cannot Create Life"

This objection sounds powerful because it contains a partial truth.

Random chance alone probably cannot create complex life in one sudden step.

But evolution does not claim that complex life arose in one step by pure random chance.

This is one of the most common misrepresentations of evolution.

Mutation and variation may occur without regard to what an organism needs. A bacterium does not decide to mutate because an antibiotic is coming. A plant does not create a useful genetic change because a drought is expected.

But natural selection is not random.

Selection filters variation. Traits that help survival or reproduction in a particular environment may become more common. Traits that harm survival or reproduction may become less common.

The process is not pure chance.

It is variation filtered by selection, inheritance, environment, and time.

The religious objection often imagines a modern cell, eye, wing, or organism suddenly assembling by accident. Then it says this is too improbable. Of course it is. But that is not what biology teaches.

Complex structures evolve through stages. Each stage can be useful in some way. Light sensitivity can precede complex vision. Gliding can precede powered flight. Simple defense can precede more complex immunity. Crude replication can precede modern genetic systems.

Evolution does not need a fully formed structure to appear all at once.

It needs useful steps.

The statement "random chance cannot create life" attacks a false version of evolution.

Science does not depend on that false version.

"The Eye Could Not Have Evolved"

The eye is one of the oldest examples used in the design argument.

It is easy to understand why.

The eye is beautiful. It receives light, focuses images, detects motion, allows reading, recognizes faces, and opens the world to the brain. The eye feels too complex to have evolved.

But feeling is not evidence.

Eyes do not need to appear suddenly in their modern form. Across life, there are many forms of light-sensitive structures. Some organisms have simple light-sensitive patches. Others have cup-shaped eyes that detect direction. Others have pinhole-like eyes. Others have lens-based eyes. Nature shows many stages of visual complexity.

The usefulness of partial vision is obvious.

A light-sensitive patch can tell an organism light from dark. That may help it move toward light, away from danger, or regulate daily rhythms. A cup-shaped structure can help detect direction. A deeper cup can improve image formation. A transparent covering can protect delicate tissue. A lens can sharpen focus.

Each step can provide some advantage.

The question "What use is half an eye?" is therefore misleading.

Half an eye can be much better than no eye.

The human eye is not perfect. The vertebrate retina has a blind spot. The eye can suffer cataract, glaucoma, retinal detachment, macular degeneration, infection, inflammation, and injury. If the eye is used as evidence of design, its weaknesses must also be counted.

Evolution explains both the eye's power and its imperfection.

Perfect divine design does not explain both as well.

"Irreducible Complexity"

Intelligent design often uses the phrase "irreducible complexity."

The claim is that some biological systems are so complex that they could not function if any part were removed. Therefore, the system could not have evolved step by step. Therefore, it must have been designed.

This sounds impressive, but it fails.

First, a system that is irreducible in its current form may still have evolved from earlier systems that had different functions. Evolution often modifies old parts for new roles. A structure does not have to begin with its final function.

Second, parts of a complex system can be added, removed, duplicated, modified, or repurposed over time. Biological systems are not like machines assembled from scratch. They are historical systems.

Third, the claim often depends on lack of imagination or lack of current knowledge. "I cannot imagine how this evolved" is not evidence that God designed it.

Fourth, intelligent design does not offer a better mechanism. Saying "an intelligent designer did it" does not explain how, when, where, by what process, or with what testable evidence.

Science answers complexity by investigation.

Intelligent design often answers complexity by surrender.

The history of science warns us against this. Many things once thought impossible to explain naturally later became understandable. The same has been true again and again in biology.

Complexity is not proof of God.

Complexity is a challenge for science, and science has repeatedly met that challenge.

“DNA Proves a Divine Programmer”

DNA is often described as a code, and this has given religious apologists a modern version of the design argument.

They say:

DNA is a code.

Every code requires a programmer.

Therefore, DNA requires God.

This sounds modern, but it is still weak.

DNA is called a code because sequences of nucleotides correspond to amino acids through cellular machinery. But scientific language should not be confused with religious metaphor. A code in biology does not automatically mean a conscious coder.

The rings of a tree contain information about age and growth conditions, but the tree was not writing a diary. A snowflake has structure, but no angel carved it. A crystal has order, but no mind arranged every atom by intention.

Natural processes can produce structure and information.

DNA also does not look like a perfect divine program.

It mutates.

It contains broken gene remnants.

It contains ancient viral insertions.

It contains duplications and repeated sequences.

It contains inherited vulnerabilities.

It can miscopy.

It can contribute to cancer and genetic disease.

If DNA is God's perfect code, why is it filled with signs of natural history?

DNA makes far more sense as an evolved molecule carrying inheritance, variation, and ancestry. Its shared patterns across species support common descent. Its errors and remnants reveal history. Its imperfections fit evolution.

Calling DNA a divine code explains nothing.

Genetics explains far more.

"The Origin of Life Proves God"

The origin of life is one of the most difficult questions in science.

How did chemistry become biology? How did the first self-sustaining, self-replicating systems arise? How did information, compartments, metabolism, and heredity come together?

Science does not yet know every step.

But that does not prove God.

This is perhaps the most important point in discussions of the origin of life.

An unanswered question is not evidence for a supernatural answer.

For centuries, religion has placed God in the unknown. But the unknown keeps shrinking. Lightning, disease, planetary motion, mental illness, heredity, species diversity, and human origins were all once explained through supernatural ideas. Science replaced those explanations with natural ones.

The origin of life is an active scientific field. Researchers study prebiotic chemistry, organic molecules, RNA, protocells, membranes, mineral surfaces, hydrothermal vents, hot springs, wet-dry cycles, energy gradients, and molecules from space.

There is no final complete answer yet.

But there is a scientific path.

"God did it" does not provide such a path. It does not explain the chemistry. It does not tell us what to test. It does not predict molecules. It does not explain how replication began. It does not explain why life uses the chemistry it uses.

The origin of life remains a frontier of science.

It is not a victory for religion.

"God Used Evolution"

Some believers accept evolution but say:

God used evolution.

This is more sophisticated than denying evolution. It allows the believer to accept much of modern science while preserving faith.

But as science, it adds nothing.

If evolution works through natural processes, and if God's guidance cannot be detected, measured, tested, or distinguished from no guidance at all, then God is not part of the scientific explanation.

A person may believe privately that God is behind evolution. That is a theological belief. But it is not a scientific conclusion.

Science asks what evidence shows.

Where is the evidence that God guided mutation?

Where is the evidence that God selected one lineage over another?

Where is the evidence that God shaped DNA beyond natural mechanisms?

Where is the evidence that God intervened in human evolution?

Where is the evidence that God prevented some extinctions and allowed others?

If no such evidence exists, then the claim "God used evolution" is not science.

It is faith added on top of science.

There is also a moral problem.

If God used evolution, then God used a process filled with suffering, disease, predation, parasitism, extinction, genetic disorders, and death. God used cancer. God used childhood disease. God used animals eating one another alive. God used mass extinction. God used billions of years of struggle before humans appeared.

That may be possible as theology, but it does not look like perfect love.

It certainly does not provide scientific evidence for God.

"Science Cannot Disprove God Completely"

This objection is more serious and must be answered carefully.

Science may not be able to disprove every possible definition of God.

If someone defines God as invisible, undetectable, beyond time, beyond space, beyond testing, and making no measurable difference in the world, then science cannot directly test such a being.

But this does not help religion as much as believers think.

A God who is completely undetectable is also scientifically unnecessary.

The question in this book is not whether every abstract god can be logically disproven. The question is whether life provides scientific evidence for God.

It does not.

There is no scientific evidence that God designed species.

There is no scientific evidence that God shaped DNA.

There is no scientific evidence that God created humans separately.

There is no scientific evidence that God guided evolution.

There is no scientific evidence that disease and extinction are part of a divine biological plan.

Science has not disproven every possible God.

But science has removed God as a necessary explanation for life.

That is enough for science-based atheism.

If God explains nothing better than natural processes, leaves no testable evidence, and makes no measurable difference, then belief in God is not a scientific conclusion.

It is faith.

And faith is not evidence.

The God of the Gaps

The God of the gaps is one of the weakest forms of belief.

It says:

Here is something we do not yet understand.

Therefore, God.

This argument has failed repeatedly.

People did not understand thunder, so they invoked gods.

They did not understand disease, so they invoked spirits.

They did not understand mental illness, so they invoked demons.

They did not understand planetary motion, so they invoked divine control.

They did not understand species, so they invoked special creation.

Science did not answer every question immediately. It often took centuries. But when answers came, they were natural.

The problem with the God of the gaps is that it makes God dependent on ignorance. The less we know, the more room God has. The more we know, the less room God has.

That is a fragile foundation for belief.

A mature view should not place God where knowledge has not yet arrived.

The unknown is not holy ground.

It is a research field.

Intelligent Design as a Religious Retreat

Intelligent design tries to appear scientific while preserving divine creation.

It avoids saying "God" directly, but the meaning is usually clear. It claims that certain features of life are best explained by an intelligent cause rather than by natural processes.

The problem is that intelligent design does not function like science.

It does not provide a productive research program comparable to evolution. It does not explain fossils better than evolution. It does not explain DNA better than evolution. It does not explain pseudogenes, endogenous retroviruses, chromosome fusion, disease, extinction, or imperfect anatomy better than evolution.

It mostly points to complexity and says, "Design."

But that is not enough.

A scientific theory must do more than express doubt about another theory. It must provide testable explanations of its own. It must make predictions. It must guide research. It must risk being wrong.

Intelligent design avoids these demands.

It often survives by criticizing evolution while offering no clear mechanism in return.

That is not science.

It is creationism in a modern suit.

Why "Design" Explains Too Much

One major weakness of design is that it can explain every possible result.

If an organ works well, it was designed.

If an organ works poorly, it was mysteriously designed.

If a species survives, God designed it to survive.

If a species goes extinct, God had another plan.

If DNA works, God wrote it.

If DNA mutates and causes disease, God's reason is unknown.

If nature is beautiful, God is loving.

If nature is cruel, God is mysterious.

This is not scientific explanation.

A scientific explanation must risk failure. It must tell us what would count against it. If nothing can count against design, then design is not being tested.

Evolution risks being wrong.

If fossils appeared in completely random order, evolution would be in trouble. If humans were found in Cambrian rocks, evolution would be in trouble. If DNA showed no meaningful pattern of relationship, evolution would be in trouble. If species were fixed and never changed, evolution would be in trouble.

But the evidence supports evolution.

Design avoids risk by explaining everything after the fact.

That is why it is scientifically weak.

The Emotional Need Behind the Objections

Many objections to evolution are not really about fossils, DNA, or natural selection.

They are about fear.

Fear that life has no purpose.

Fear that humans are only animals.

Fear that morality will collapse.

Fear that death is final.

Fear that suffering has no divine meaning.

Fear that childhood beliefs were wrong.

Fear that family, culture, and community may be shaken.

These fears are human. They deserve understanding.

But fear does not make a belief true.

A person may need God emotionally. That does not provide scientific evidence for God. A person may feel comforted by divine design. That does not mean life was designed. A person may feel lost without special creation. That does not make special creation true.

Truth must be approached with courage.

Science-based atheism does not say life is worthless. It says life is natural. It does not say morality is impossible. It says morality is human responsibility. It does not say wonder is dead. It says wonder does not require superstition.

The emotional need for God explains why belief persists.

It does not prove God exists.

The Difference Between Possibility and Evidence

A common mistake is to confuse possibility with evidence.

Is it possible that some hidden God exists beyond all testing? A person may say yes.

But possibility is not evidence.

Many things are possible in the abstract. It may be possible that unknown beings exist somewhere in the universe. It may be possible that reality contains dimensions we do not understand. It may be possible that some intelligence beyond human imagination exists in some form.

But science does not accept claims merely because they are possible.

Science asks for evidence.

The claim "God designed life" is not supported by the evidence of biology. Life shows natural ancestry, mutation, selection, disease, extinction, and imperfection. These are not the marks of perfect divine design. They are the marks of evolution.

A God may be imagined.

But imagination is not evidence.

Faith may assert.

But assertion is not evidence.

Science requires more.

What Would Change Our Mind?

A serious scientific view must be open to correction.

So what kind of evidence would challenge evolution or support design?

If fossils appeared in completely impossible sequences, that would matter. If complex modern mammals appeared in the earliest fossil layers, that would matter. If DNA showed no pattern of common ancestry, that would matter. If species were found to be fixed and unable to change, that would matter. If clear, testable, repeatable evidence of divine intervention in biological systems were found, that would matter.

Science is not closed to evidence.

But the evidence we have does not point toward design.

It points toward evolution.

This is the difference between science and dogma. Science can state what would count against it. Religion often cannot. The believer can always say God works mysteriously. But if every observation fits the belief, the belief is insulated from evidence.

Insulated beliefs may be comforting.

They are not scientific.

Why Natural Explanation Is Enough

At this stage, the central conclusion of the book becomes clear.

We do not need God to explain why life looks designed. Natural selection explains adaptation.

We do not need God to explain the fossil record. Deep time and evolution explain it.

We do not need God to explain DNA. Genetics and common ancestry explain it.

We do not need God to explain the tree of life. Descent with modification explains it.

We do not need God to explain imperfect bodies. Evolutionary tinkering explains them.

We do not need God to explain disease. Mutation, infection, immune dysfunction, aging, trade-offs, and biological limits explain it.

We do not need God to explain parasites and predation. Evolutionary competition explains them.

We do not need God to explain extinction. Changing environments, catastrophe, adaptation, and chance explain it.

We do not need God to explain human origins. Human evolution explains them.

At each stage, the natural explanation is stronger, clearer, and more useful than divine design.

God adds no scientific necessity.

Belief After Evidence

Some people will continue to believe in God after accepting all of this.

That is their right.

But the nature of the belief has changed.

It can no longer honestly claim that science requires God to explain life. It can no longer claim that species must have been specially created. It can no longer claim that the body is obvious proof of perfect design. It can no longer claim that humans are biologically separate from other animals. It can no longer claim that unknown details are proof of divine action.

The belief becomes faith.

Private faith may remain meaningful to some people. But faith is not scientific evidence.

The purpose of this book is not to control private emotion. It is to examine public claims about reality.

And the claim that life scientifically proves God does not survive careful examination.

The Last Retreat: Mystery

At the end, religious belief often retreats to mystery.

God is beyond science.

God's ways are beyond human understanding.

God works invisibly.

God does not need to leave evidence.

God is not testable.

This is a very different claim from the original design argument.

The design argument says life shows evidence of God.

The mystery argument says God does not need evidence.

Those are not the same.

If a believer says God is beyond evidence, then the believer should not also claim that life is evidence for God. One cannot have it both ways.

Either life provides scientific evidence of divine design, or God is beyond scientific evidence.

This book has examined the first claim.

It fails.

Life does not provide scientific evidence of God.

If someone continues to believe without evidence, that is faith. But it should be honestly named as faith, not science.

Science-Based Atheism and Intellectual Honesty

Science-based atheism does not require hatred of religion.

It requires intellectual honesty.

It says we should not call mystery evidence.

We should not call complexity proof.

We should not call tradition truth.

We should not call emotional comfort science.

We should not call ignorance God.

It asks us to follow evidence even when evidence leads away from childhood belief, cultural habit, family tradition, or religious authority.

This can be difficult.

But growing intellectually often requires leaving behind explanations that no longer fit reality.

A child may think the moon follows the car. Later, the child learns astronomy. A patient may think antibiotics cure every infection. Later, the patient learns about viruses and resistance. A society may think disease is punishment. Later, it learns microbiology.

Human understanding matures by replacing comforting errors with better explanations.

The same must happen with divine design.

Life does not need God as an explanation.

Science is enough.

The Calm Conclusion

The designer's last hiding places do not hold.

"Evolution is only a theory" misunderstands science.

"Microevolution but not macroevolution" creates an artificial wall.

"There are no transitional fossils" is false.

"Random chance cannot create life" misrepresents evolution.

"The eye could not evolve" ignores useful intermediate stages.

"Irreducible complexity" mistakes current complexity for impossible history.

"DNA proves a divine programmer" turns a metaphor into a false argument.

"The origin of life proves God" confuses ignorance with evidence.

"God used evolution" is theology, not science.

"Science cannot disprove God completely" does not show that life provides evidence for God.

Again and again, the same pattern appears.

When science explains more, God retreats. When evidence grows, divine design becomes less necessary. When mechanisms are discovered, mystery loses its power as proof.

The living world does not need a hidden designer to make sense.

It needs natural processes.

Evolution, genetics, deep time, chemistry, selection, disease, extinction, and common ancestry explain life better than God.

The conclusion is not angry.

It is clear.

There is no scientific evidence that life was designed by God.

The designer survives only by becoming invisible, undetectable, untestable, and unnecessary.

That is not a scientific explanation.

It is the end of one.

Chapter Summary

Religious belief often retreats into gaps when science explains what was once attributed to God.

The phrase "evolution is only a theory" misunderstands the meaning of scientific theory.

Microevolution and macroevolution are not separate miracles. Larger evolutionary patterns can arise from accumulated changes, speciation, extinction, and deep time.

Transitional fossils exist and strongly support evolution.

Evolution is not pure random chance. Variation may arise without regard to need, but natural selection filters variation.

Complex structures such as the eye can evolve through useful intermediate stages.

Irreducible complexity does not prove design because biological systems can evolve through modification, repurposing, duplication, and changing function.

DNA does not prove a divine programmer. It shows inheritance, mutation, ancestry, and imperfection.

The origin of life remains an active scientific question, not evidence for God.

The claim that God used evolution is a theological belief, not a scientific explanation.

Science may not disprove every abstract God, but it has removed God as a necessary explanation for life.

Key Takeaway

The last hiding places of the designer are not scientific shelters. They are retreats from evidence. Evolution is well supported, DNA reveals ancestry, fossils show transition, disease exposes imperfection, extinction reveals natural history, and human evolution removes special creation. A God who leaves no testable evidence and explains nothing better than nature is not a scientific explanation. Life does not provide scientific evidence for God. No designer is required.

Sources

Charles Darwin. On the Origin of Species.

National Academies of Sciences, Engineering, and Medicine. Science, Evolution, and Creationism.

National Academies of Sciences, Engineering, and Medicine. Evolution Resources.

American Association for the Advancement of Science. Statement on Intelligent Design.

National Center for Science Education. Resources on Evolution, Creationism, and Intelligent Design.

Kitzmiller v. Dover Area School District, 2005.

University of California Museum of Paleontology. Understanding Evolution: Misconceptions About Evolution.

University of California Museum of Paleontology. Understanding Evolution: Transitional Features.

Jerry A. Coyne. Why Evolution Is True.

Richard Dawkins. The Blind Watchmaker.

Kenneth R. Miller. Finding Darwin's God.

Barbara Forrest and Paul R. Gross. Creationism's Trojan Horse.

Transition to the Conclusion

The argument is now complete.

Life looks designed, but natural selection explains that appearance. Fossils show deep time and change. DNA reveals common ancestry. The tree of life connects all living beings. Imperfect bodies show tinkering, not engineering. Disease, parasites, suffering, and extinction challenge the idea of perfect divine design. Human evolution removes special creation. Intelligent design and God of the gaps fail as science.

The conclusion will now bring these lines together.

Life is beautiful.

Life is fragile.

Life is meaningful to us.

But life is not evidence of God.

Life is evolved, not designed.

Conclusion: Life Without a Designer, Meaning Without God

The Beautiful Truth

Life is beautiful.

A newborn taking the first breath. A bird lifting into the morning sky. A flower opening after rain. A child learning to speak. A wound slowly healing. A mother holding her baby. A physician listening to a heartbeat. A forest alive with sound. A human mind asking where it came from.

No honest person should look at life with indifference.

Life deserves wonder.

But wonder is not evidence of God.

That has been the central message of this book. Living beings can move us deeply. They can fill us with awe, tenderness, curiosity, gratitude, and humility. But the emotional power of life does not turn religion into science.

When life is viewed from a distance, it may look designed.

When life is examined closely, with biology, genetics, medicine, fossils, and evolution, a different picture appears.

Life is not a perfect machine.

Life is not a collection of separate divine products.

Life is not centered on human beings.

Life is not protected by divine mercy.

Life is not free from disease, waste, suffering, or extinction.

Life is evolved.

That truth is not cold.

It is clear.

Appearance Is Not Proof

The design argument begins with appearance.

The eye looks made for seeing. The wing looks made for flying. The heart looks made for pumping blood. The flower looks made for the bee. The human body looks made for life.

This appearance is powerful. It is one of the reasons belief in divine design has remained so attractive to the human mind.

But appearance is not proof.

The sun appears to move around the Earth, but it does not. The Earth appears flat from the ground, but it is not.

Disease once appeared to come from spirits, curses, or divine punishment, but it does not.

Biology has shown that the appearance of design can arise without a designer.

Natural selection explains why useful traits become more common. Variation supplies possibilities. Inheritance carries them forward. Survival and reproduction filter them. Deep time allows small changes to become large differences.

Life looks designed because natural selection can create adaptation.

That insight changed the human understanding of life forever.

It removed the divine designer from the center of biology.

Darwin Changed the Question

Before Darwin, divine creation felt reasonable to many people.

They did not know DNA. They did not understand genes. They did not know the true age of the Earth. They did not understand extinction properly. They did not see life as a branching tree. They did not know how adaptation could arise through natural causes.

So God filled the gap.

God created species.

God designed organs.

God made human beings separately.

God gave life its purpose.

That explanation felt complete in a world without modern biology.

But science changed the world.

Geology revealed deep time. Fossils revealed vanished life. Darwin revealed natural selection. Genetics revealed inheritance. DNA revealed common ancestry. Medicine revealed the body's natural vulnerability. Evolutionary biology revealed that living beings are not fixed products, but historical populations.

The old creation story did not survive as science.

It may still survive as faith, tradition, poetry, culture, and emotional comfort.

But it no longer survives as a scientific explanation.

DNA Revealed the Family Archive

Darwin did not know DNA.

That makes the modern evidence even more powerful.

DNA has confirmed and strengthened the central idea of common ancestry. It shows inheritance, mutation, shared genes, damaged gene remnants, ancient viral traces, chromosomal clues, and branching patterns of relationship.

If living beings were separately created, we might expect separate biological histories.

Instead, we find kinship.

Humans share deep molecular connections with other life. We are mammals. We are primates. We are great apes. Our genome carries signs of ancestry, not signs of separate creation.

DNA does not say, "God made each creature separately."

DNA says, "Life is related."

The genome is not a perfect message from heaven.

It is a biological archive.

The Tree of Life Replaced Separate Creation

The living world is not a showroom of divine designs.

It is a tree.

Bacteria, plants, fungi, insects, fish, reptiles, birds, mammals, apes, and humans belong to a vast history of descent, branching, modification, survival, and extinction.

This truth is humbling.

Human beings are not above the tree.

We are part of it.

Religion often placed humans at the center of creation. Science places us within nature. This is not a loss of dignity. It is a gain in honesty.

We are not separate creations.

We are relatives of life.

That kinship should not make us arrogant. It should make us responsible.

The Body Shows History, Not Perfect Engineering

The human body is remarkable, but it is not perfectly designed.

The spine supports upright walking, but it often causes pain. The throat allows speech, but it creates a risk of choking. The eye sees, but it has a blind spot. The recurrent laryngeal nerve takes a long detour. Wisdom teeth often do not fit. The appendix can become inflamed and life-threatening. Childbirth can be dangerous. The immune system can attack the body. Cells that divide can become cancer.

These are not the signs of flawless divine engineering.

They are the signs of evolutionary tinkering.

Evolution does not start with a clean blueprint. It modifies old structures. It works with inherited material. It keeps what works well enough. It cannot plan for every future. It cannot remove every flaw.

The body works, but it works as history.

It is not a machine built from heaven.

It is a living record of ancestry, compromise, and survival.

Disease Is a Deep Wound in the Design Argument

Disease is one of the strongest arguments against perfect divine design.

The body can heal, but it can also fail. DNA carries inheritance, but it can mutate. Cell division allows growth, but it can produce cancer. The immune system protects, but it can become autoimmune. Reproduction creates life, but it can bring miscarriage, birth defects, dangerous pregnancy, and maternal death. The brain allows thought, but it can suffer mental illness, seizures, dementia, and degeneration.

Infectious disease shows life invading life. Parasites exploit bodies. Viruses use cells. Bacteria evolve resistance. Pain protects, but it can become torture. Aging slowly takes away strength, memory, vision, mobility, and independence.

A perfect, loving, all-powerful designer could have made life differently.

Evolution explains why life is this way.

Disease arises because biology is natural, imperfect, and historical. Bodies are shaped by mutation, trade-offs, infection, aging, developmental risk, immune balance, and environmental mismatch.

Disease does not prove a divine plan.

It exposes the weakness of that claim.

Nature Is Beautiful, But Not Merciful

Beyond human disease, nature itself challenges divine design.

Nature is beautiful, but it is not merciful.

Predators kill. Parasites invade. Viruses hijack cells. Young animals starve. Offspring are produced in vast numbers so that a few may survive. Animals suffer injuries, infections, hunger, fear, and slow death.

This suffering existed long before human beings.

It cannot be explained scientifically as punishment for human sin. It cannot be blamed on human disobedience. It cannot be placed inside a simple moral drama.

Nature was already full of suffering before humans arrived.

Evolution explains this. Life survives by using energy, reproducing, competing, adapting, cooperating, and sometimes exploiting other life. It has no moral intention. It does not promise fairness. It does not protect innocence.

Nature is not cruel because it has evil intentions.

Nature is indifferent because it has no intentions.

That is not the world we would expect from perfect divine love.

It is the world we would expect from evolution.

Extinction Shows No Divine Plan

Extinction is the graveyard beneath every design argument.

Most species that ever lived are gone. Trilobites, ammonites, non-avian dinosaurs, mammoths, saber-toothed cats, and many human relatives vanished. Entire ecosystems flourished and disappeared. Mass extinctions erased large portions of life.

Extinction is not rare.

It is central to life's history.

This creates a serious problem for divine design.

Why would a perfect creator make countless species only to erase most of them? Why create entire worlds of life before humans existed? Why allow complex, beautiful, successful organisms to vanish forever?

Evolution has no difficulty with extinction.

Species survive only as long as conditions allow. When environments change beyond what a lineage can endure, extinction follows. There is no moral guarantee. No sacred protection. No divine preservation of beauty or innocence.

The fossil record is not a record of perfect planning.

It is a record of life trying, changing, surviving, and disappearing.

Human Beings Were Not Specially Created

Human evolution is one of the most important truths in science.

We did not evolve from modern chimpanzees. We share common ancestors with them. We are mammals, primates, and great apes. Our bodies carry the marks of ancestry. Our fossils show a branching human family. Our DNA connects us to other life. Neanderthals, Denisovans, and other human relatives show that human history was not a simple story of one created pair.

Human beings were not inserted into nature.

We grew out of nature.

This truth directly challenges special creation. Religion often says humans are made in the image of God. Biology says humans are evolved animals with extraordinary minds. Religion says we were placed above nature. Biology says we are part of nature.

This does not destroy human dignity.

It removes supernatural pride.

We do not need divine creation to matter. We matter because we are conscious, vulnerable, social beings. We can suffer. We can love. We can remember. We can reason. We can care. We can build medicine, science, law, art, friendship, and justice.

Human dignity does not depend on a soul.

It depends on the reality of lived experience.

Faith Is Not Evidence

Some readers may still believe in God after reading this book.

That is their personal right.

But the belief should be named honestly.

It is faith.

It is not scientific evidence.

There is no scientific evidence that God designed life. There is no scientific evidence that God created species separately. There is no scientific evidence that God shaped DNA. There is no scientific evidence that God made humans outside evolution. There is no scientific evidence that disease, parasites, predation, and extinction are part of a divine biological plan.

A person may believe without scientific evidence.

But belief without scientific evidence should not be presented as science.

This distinction is essential.

The purpose of this book has not been to insult believers. It has been to examine a claim. The claim is that life proves God. When we examine the evidence carefully, that claim fails.

Life does not prove God.

Life reveals the power of nature.

Meaning Does Not Need God

Many people fear that if life is not designed, then life has no meaning.

This fear is understandable, but false.

A life does not need to be divinely planned to be meaningful.

A child's laughter has meaning. A mother's love has meaning. A doctor's care has meaning. A friendship has meaning. A just law has meaning. A poem has meaning. A meal shared with family has meaning. A hand held at the bedside has meaning. A mind searching honestly for truth has meaning.

Meaning is not handed down from the sky.

Meaning is created, discovered, shared, and lived by conscious beings.

In fact, a natural view of life can make meaning more urgent. If this life is not a rehearsal for another world, then this life matters more. If no divine plan guarantees justice, then justice must be built by human beings. If no god prevents suffering, then compassion becomes our responsibility. If no heavenly designer protects the living world, then we must protect it ourselves.

Atheism is not emptiness.

It is responsibility without illusion.

Wonder Without Worship

Science does not kill wonder.

It purifies wonder.

It removes superstition and leaves reality more visible. The rainbow is not less beautiful because we understand light. The flower is not less beautiful because we understand pollination. The eye is not less remarkable because we understand evolution. The stars are not less magnificent because we understand physics. Human love is not less real because the brain and body are involved.

Understanding does not make life smaller.

It makes life deeper.

The old religious answer says, “God made it.”

Science asks, “How did it come to be?”

That second question opens the universe.

It leads to fossils, DNA, cells, stars, chemistry, evolution, medicine, and the human mind itself. It teaches humility because we are not the center. It teaches kinship because we are related to life. It teaches responsibility because there is no divine parent coming to rescue us from our own ignorance.

This is not a cold view.

It is a mature view.

It is wonder without worship.

The Final Scientific Conclusion

The evidence in this book leads to one conclusion:

There is no scientific evidence that life was designed by God.

Life is not a perfect machine.

Life is not a collection of separate divine products.

Life is not centered on human beings.

Life is not protected by divine mercy.

Life is not free from waste, suffering, disease, or extinction.

Life is not explained better by God than by nature.

Life is evolved.

It emerged from chemistry, developed through inheritance, changed through variation, was shaped by selection, recorded itself in DNA, branched through common ancestry, suffered through disease and predation, survived through adaptation, and lost most of its forms to extinction.

This is not a story of divine design.

It is a story of natural history.

And natural history is enough.

Transition to the Next Book

This book has shown that life does not need a designer.

But one of religion's deepest claims still remains.

Even if the body evolved, what about the mind?

Even if life does not need God, what about consciousness?

Even if humans are animals, what about morality?

Even if the brain is biological, what about the soul?

These questions will guide the next book in the No God Required series.

Book 5 will move from life to mind.

It will examine whether consciousness requires a soul, whether morality requires God, and whether the human mind can be understood through the brain, evolution, social life, and reason.

The journey will continue from mystery to evidence, from supernatural belief to natural explanation, from religious comfort to scientific honesty.

From God as explanation to reality as it is.

Final Summary

Life is beautiful, but beauty is not evidence of God.

Life looks designed, but natural selection explains the appearance of design.

Darwin gave biology a natural mechanism for adaptation and diversity.

Evolution is observable in real time through bacteria, viruses, plants, insects, and changing populations.

Deep time and fossils show that life changed across Earth's long history.

The origin of life remains an active scientific question, not proof of God.

DNA reveals inheritance, mutation, shared ancestry, and the historical archive of life.

The tree of life shows that living beings are one family, not separate creations.

The body shows tinkering, compromise, and inherited structure, not perfect engineering.

Disease exposes the weakness of the claim that the body was designed by perfect love.

Parasites, predation, starvation, and waste show that nature is indifferent, not merciful.

Extinction reveals dead branches, vanished worlds, and no smooth divine plan.

Human evolution removes special creation while preserving natural dignity.

Intelligent design and God of the gaps fail as science.

There is no scientific evidence that God designed life.

Final Takeaway

Life does not need a designer to be beautiful. Human beings do not need special creation to have dignity. The living world does not show the signature of God. It shows the power of nature, the depth of time, the reality of evolution, and the fragile wonder of existence.

God is not required.

Sources

Charles Darwin. On the Origin of Species.

Charles Darwin. The Descent of Man.

National Academies of Sciences, Engineering, and Medicine. Science, Evolution, and Creationism.

University of California Museum of Paleontology. Understanding Evolution.

National Human Genome Research Institute. Genetics and Genomics Resources.

Smithsonian National Museum of Natural History. Human Origins Program.

NASA Astrobiology. Origin-of-Life and Prebiotic Chemistry Resources.

Randolph M. Nesse and George C. Williams. Why We Get Sick.

Jerry A. Coyne. Why Evolution Is True.

Richard Dawkins. The Blind Watchmaker.

Neil Shubin. Your Inner Fish.

Donald R. Prothero. Evolution: What the Fossils Say and Why It Matters.

Carl Zimmer. Evolution: The Triumph of an Idea.

Appendix: Main Sources and Further Reading

Why This Appendix Matters

This book has made a strong claim:

Life does not provide scientific evidence for God.

That claim should not rest on anger, emotion, or personal dislike of religion. It should rest on evidence. The purpose of this appendix is to give readers a clear path to the major scientific sources behind the ideas discussed in this book.

This is not meant to be an exhausting academic bibliography. A list that is too long may look impressive, but it can also overwhelm ordinary readers. The aim here is practical. This appendix provides a focused, reliable, and readable guide for readers who want to examine the scientific foundation of evolution, natural selection, DNA, fossils, human origins, disease, extinction, and intelligent design.

The central point of the book is simple:

Modern biology does not need divine design to explain life.

The sources below help show why.

How to Use This Appendix

Readers do not need to read every source listed here.

A general reader may begin with the "Recommended Reading Path for General Readers." A reader interested in the medical side may begin with evolutionary medicine. A reader interested in human origins may begin with the Smithsonian Human Origins Program and books on Neanderthals, Denisovans, and early humans. A reader interested in creationism and intelligent design may begin with the National Academies, the National Center for Science Education, and the Dover case.

The important thing is not to memorize every source.

The important thing is to see that the argument in this book does not depend on one scientist, one fossil, one gene, one laboratory, or one opinion.

It rests on convergence.

Fossils, DNA, anatomy, embryology, medicine, observed evolution, and comparative biology all point toward the same conclusion:

Life is evolved, not separately created.

Recommended Reading Path for General Readers

For readers who want a clear and accessible start, these sources are the best place to begin.

Jerry A. Coyne. Why Evolution Is True.

Neil Shubin. Your Inner Fish.

Richard Dawkins. The Blind Watchmaker.

Carl Zimmer. Evolution: The Triumph of an Idea.

University of California Museum of Paleontology. Understanding Evolution.

National Academies of Sciences, Engineering, and Medicine. Science, Evolution, and Creationism.

These sources explain evolution, common ancestry, fossils, anatomy, DNA, and the failure of creationist objections in a way that is suitable for thoughtful non-specialist readers.

Evolution and Natural Selection

Evolution is the central scientific explanation for the diversity and relatedness of life. These sources are useful for understanding natural selection, adaptation, descent with modification, and the appearance of design without a designer.

Charles Darwin. On the Origin of Species.

Darwin's central work remains one of the most important books in the history of science. It introduced natural selection as a powerful explanation for adaptation, diversity, and descent with modification. The language is older, but the central insight remains foundational.

Charles Darwin. The Descent of Man.

This book applied evolutionary thinking to human beings. It is historically important because it directly challenged the idea that humans were specially created outside nature.

Theodosius Dobzhansky. Nothing in Biology Makes Sense Except in the Light of Evolution.

This famous essay explains why evolution is the central organizing idea of biology. It remains one of the clearest short statements on the importance of evolution.

Ernst Mayr. What Evolution Is.

A clear explanation of evolution by one of the major evolutionary biologists of the twentieth century. Useful for understanding species, variation, selection, and common ancestry.

Jerry A. Coyne. Why Evolution Is True.

A strong and readable book for general readers. It explains fossils, DNA, anatomy, biogeography, and other evidence for evolution in clear language.

Richard Dawkins. The Blind Watchmaker.

A classic explanation of how natural selection can produce the appearance of design without a designer. This book is especially relevant to the main argument of this book.

Richard Dawkins. The Greatest Show on Earth.

A broad presentation of evidence for evolution, including fossils, genetics, embryology, and observed evolution.

Sean B. Carroll. Endless Forms Most Beautiful.

A readable introduction to evolutionary developmental biology. It helps explain how changes in development can contribute to the diversity of living forms.

Fossils, Deep Time, and Transitional Forms

The fossil record shows that life has a long history. Species appeared, changed, branched, and disappeared across deep time. These sources help readers understand why the rocks support evolution, not sudden divine creation.

Charles Lyell. Principles of Geology.

A foundational work in geology that helped establish the importance of deep time. Deep time was essential for making gradual evolution understandable.

Donald R. Prothero. Evolution: What the Fossils Say and Why It Matters.

A detailed and useful source on fossil evidence for evolution. It also addresses many common creationist objections.

Neil Shubin. Your Inner Fish.

A highly readable book showing how the human body carries evidence of ancient ancestry. It connects anatomy, fossils, and evolution in a memorable way.

Stephen Jay Gould. Wonderful Life.

A famous book on the Burgess Shale and the complexity of life's history. Useful for understanding contingency, extinction, and the branching nature of evolution.

Carl Zimmer. Evolution: The Triumph of an Idea.

A broad and engaging overview of evolutionary science, suitable for general readers.

University of California Museum of Paleontology. Understanding Evolution.

A reliable educational resource explaining natural selection, adaptation, fossils, common ancestry, speciation, and common misconceptions about evolution.

National Academies of Sciences, Engineering, and Medicine. Science, Evolution, and Creationism.

A concise and authoritative source explaining the scientific evidence for evolution and the difference between science and religious belief.

Genetics, DNA, and Common Ancestry

DNA has made the evidence for evolution even stronger. It shows inheritance, mutation, shared ancestry, broken gene remnants, ancient viral traces, and molecular kinship across life.

National Human Genome Research Institute. Genetics and Genomics Resources.

A reliable source for understanding genes, DNA, mutations, inherited disorders, and modern genomics.

National Human Genome Research Institute. Gene.

A helpful reference explaining what genes are and how they relate to inheritance and biological traits.

National Human Genome Research Institute. Mutation.

A clear explanation of mutation as a change in DNA sequence and how such changes can arise.

National Human Genome Research Institute. Comparative Genomics.

A useful source for understanding how comparison of genomes reveals similarities and differences among species.

National Human Genome Research Institute. Genetic Disorders.

A reliable resource for understanding inherited disease and the medical importance of genetic variation.

National Academies Press. Evidence Supporting Biological Evolution.

A useful discussion of multiple lines of evidence for evolution, including molecular evidence and shared genetic features.

Sean B. Carroll. The Making of the Fittest.

A readable book showing how DNA records evolution and how genetics supports natural selection and common ancestry.

J. W. IJdo and colleagues. Origin of Human Chromosome 2: An Ancestral Telomere-Telomere Fusion.

An important scientific paper on the evidence that human chromosome 2 resulted from an ancestral chromosome fusion.

W. E. Johnson. Origins and Evolutionary Consequences of Ancient Endogenous Retroviruses.

A useful source for understanding ancient viral remnants in genomes and their evolutionary significance.

The Tree of Life and LUCA

The tree of life shows that living beings are not separate divine creations. They are related through deep ancestry. These sources help explain the unity and complexity of life's relationships.

University of California Museum of Paleontology. Understanding Evolution: The Tree of Life.

A clear educational source explaining how living beings are related through branching ancestry.

University of California Museum of Paleontology. Understanding Evolution: Homologies and Analogies.

A helpful resource for understanding shared structures due to common ancestry and similar traits due to similar selection pressures.

W. Ford Doolittle. Uprooting the Tree of Life.

An important paper discussing how horizontal gene transfer complicates the simple tree image, especially among microbes.

W. Ford Doolittle and Eric Bapteste. Pattern Pluralism and the Tree of Life Hypothesis.

A more technical discussion of the complexity of life's deepest relationships.

M. C. Weiss and colleagues. The Last Universal Common Ancestor Between Ancient Earth Chemistry and the Onset of Genetics.

A scientific discussion of LUCA, the Last Universal Common Ancestor, and its place in early life history.

Nick Lane. The Vital Question.

A thoughtful book connecting energy, cells, and the deep origin of complex life. Useful for readers who want a deeper look at early life and cellular evolution.

Origin of Life and Abiogenesis

The origin of life is still an open scientific question. That does not make it evidence for God. These sources show how science investigates the transition from chemistry to biology.

NASA Astrobiology. Origin-of-Life Resources.

NASA's astrobiology material provides reliable information on prebiotic chemistry, early Earth environments, organic molecules, and the search for life beyond Earth.

NASA. Asteroid and Meteorite Sample Research on Organic Molecules.

Useful for readers interested in how organic molecules can form naturally in space and may have been delivered to early Earth.

Bruce Alberts and colleagues. Molecular Biology of the Cell.

A major biology textbook. Relevant sections explain RNA, cells, molecular biology, and the possible importance of RNA in early life.

David Deamer and Bruce Damer. The Hot Spring Hypothesis for an Origin of Life.

A useful source for readers interested in the possibility that early life emerged in hot spring or wet-dry cycling environments.

Addy Pross. What Is Life? How Chemistry Becomes Biology.

A thoughtful book on the transition from chemistry to biology.

John Maynard Smith and Eörs Szathmáry. The Major Transitions in Evolution.

A deeper scientific work on major evolutionary transitions, including the emergence of genetic systems, cells, and complex life.

J. Gómez-Márquez. The Origin of Life and Cellular Systems.

A useful scientific review for readers who want a more technical look at origin-of-life questions.

Nick Lane. The Vital Question.

Relevant again for understanding energy, early life, and the evolution of complex cells.

Evolutionary Medicine and Disease

Disease is central to the argument against perfect divine design. Evolutionary medicine helps explain why bodies are powerful but vulnerable, and why disease is not a mystery requiring supernatural explanation.

Randolph M. Nesse and George C. Williams. Why We Get Sick: The New Science of Darwinian Medicine.

A foundational book in evolutionary medicine. It explains why bodies are vulnerable to disease and how evolution helps us understand illness.

Randolph M. Nesse. Good Reasons for Bad Feelings.

A thoughtful book on evolution, emotion, anxiety, depression, and mental health.

Peter D. Gluckman, Alan Beedle, and Mark Hanson. Principles of Evolutionary Medicine.

A more academic but very useful source on evolutionary explanations for disease vulnerability.

Stephen C. Stearns and Jacob C. Koella. Evolution in Health and Disease.

A strong academic source for readers interested in the medical implications of evolution.

M. L. Benton and colleagues. The Influence of Evolutionary History on Human Health and Disease.

A useful scientific discussion of how evolutionary history affects human disease risk.

National Cancer Institute. What Is Cancer?

A reliable source explaining cancer as uncontrolled cell growth and spread.

Centers for Disease Control and Prevention. Antimicrobial Resistance Resources.

Useful for understanding antibiotic resistance as a major public health problem and as a clear example of evolution in real time.

World Health Organization. Antimicrobial Resistance Resources.

A reliable global public health source on antimicrobial resistance.

National Human Genome Research Institute. Genetic Disorders.

Useful for understanding inherited disease and the medical importance of genetic variation.

Human Evolution

Human evolution directly challenges special creation. These sources help explain the fossil, anatomical, genetic, and ancient DNA evidence showing that humans are part of the natural history of life.

Smithsonian National Museum of Natural History. Human Origins Program.

One of the best public educational resources on human evolution. It covers fossils, tools, climate, behavior, and ancestry.

Charles Darwin. The Descent of Man.

Historically important for understanding Darwin's application of evolution to human beings.

Richard Leakey. The Origin of Humankind.

A readable introduction to human evolution and the fossil record.

Donald Johanson and Maitland Edey. Lucy: The Beginnings of Humankind.

A classic account of the discovery and significance of Lucy and early human ancestors.

Chris Stringer. Lone Survivors: How We Came to Be the Only Humans on Earth.

A useful book on the evolution and survival of Homo sapiens.

Svante Pääbo. Neanderthal Man: In Search of Lost Genomes.

A fascinating account of ancient DNA, Neanderthals, Denisovans, and human ancestry.

Ian Tattersall. Masters of the Planet.

A broad account of human evolution, cognition, and the development of modern humans.

National Human Genome Research Institute. Human Chromosome 2 Resources.

Useful for understanding the chromosome fusion evidence that supports common ancestry between humans and other great apes.

Intelligent Design, Creationism, and Science

The last chapters of this book examine why intelligent design and God of the gaps fail as science. These sources help explain the difference between scientific explanation and religious assertion.

National Academies of Sciences, Engineering, and Medicine. Science, Evolution, and Creationism.

An important source for understanding why evolution is science and why creationism and intelligent design do not function as scientific alternatives.

American Association for the Advancement of Science. Statement on Intelligent Design.

A clear statement explaining why intelligent design has not met the standard of science.

National Center for Science Education. Evolution, Creationism, and Intelligent Design Resources.

A useful source for understanding common creationist arguments and scientific responses.

Kitzmiller v. Dover Area School District, 2005.

A major legal case in the United States that examined intelligent design in public school science education. It is important historically because it showed that intelligent design was closely connected to creationism and did not function as science.

Barbara Forrest and Paul R. Gross. Creationism's Trojan Horse.

A detailed critical examination of the intelligent design movement.

Kenneth R. Miller. Finding Darwin's God.

Written by a religious scientist who accepts evolution. Useful for understanding how some believers reconcile faith and evolution, although the conclusion of this book differs from Miller's theological position.

Richard Dawkins. The Blind Watchmaker.

Relevant again because it directly answers the claim that biological complexity requires divine design.

Jerry A. Coyne. Why Evolution Is True.

Relevant again because it responds to many creationist objections in clear language.

Nature, Suffering, and Natural Evil

Nature is beautiful, but it is not merciful. These sources help readers understand adaptation, natural suffering, parasitism, predation, extinction, and the weakness of the claim that nature clearly reveals divine love.

Charles Darwin. Letter to Asa Gray, May 22, 1860.

Darwin's reflection on parasitic wasps is important because it shows how cruelty in nature challenged simple belief in a benevolent designer.

George C. Williams. Adaptation and Natural Selection.

A serious evolutionary work explaining adaptation without romanticizing nature.

David Attenborough. Life on Earth.

A classic work presenting the diversity and history of life in vivid form.

Richard Dawkins. River Out of Eden.

A short and powerful book discussing evolution, indifference, and the natural world without divine purpose.

Randolph M. Nesse and George C. Williams. Why We Get Sick.

Relevant again for understanding suffering and disease through evolution rather than divine design.

Stephen Jay Gould. Wonderful Life.

Relevant again for understanding contingency, extinction, and the lack of a simple human-centered plan in life's history.

University of California Museum of Paleontology. Understanding Evolution: Coevolution.

Useful for understanding predator-prey relationships, host-parasite interactions, competition, and mutualism as natural evolutionary processes.

Extinction and Mass Extinction

Extinction is one of the strongest challenges to perfect divine design. These sources explain why extinction is not rare, why most species are gone, and why the fossil record is a history of both life and disappearance.

University of California Museum of Paleontology. Understanding Evolution: Extinction.

A clear educational source explaining extinction as a major part of evolutionary history.

University of California Museum of Paleontology. Understanding Evolution: Mass Extinction.

A useful introduction to major extinction events and their role in reshaping life.

Douglas H. Erwin. Extinction: How Life on Earth Nearly Ended 250 Million Years Ago.

A detailed account of the Permian-Triassic extinction, one of the most severe mass extinctions known.

Peter Ward. Under a Green Sky.

A readable discussion of extinction, climate, ocean change, and life's vulnerability.

Donald R. Prothero. Evolution: What the Fossils Say and Why It Matters.

Useful again for understanding fossil evidence, extinction, and creationist objections.

Natural History Museum. What Killed the Dinosaurs?

A helpful public resource on the extinction of non-avian dinosaurs and the Cretaceous-Paleogene event.

Recommended Reading Path by Topic

Readers interested in **evolution and natural selection** may begin with:

Jerry A. Coyne. Why Evolution Is True.

Richard Dawkins. The Blind Watchmaker.

University of California Museum of Paleontology. Understanding Evolution.

National Academies of Sciences, Engineering, and Medicine. Science, Evolution, and Creationism.

Readers interested in **fossils and deep time** may begin with:

Neil Shubin. Your Inner Fish.

Donald R. Prothero. Evolution: What the Fossils Say and Why It Matters.

Carl Zimmer. Evolution: The Triumph of an Idea.

University of California Museum of Paleontology. Fossil Evidence Resources.

Readers interested in **DNA and common ancestry** may begin with:

National Human Genome Research Institute. Comparative Genomics.

Sean B. Carroll. The Making of the Fittest.

Jerry A. Coyne. Why Evolution Is True.

Richard Dawkins. The Greatest Show on Earth.

Readers interested in **disease and medicine** may begin with:

Randolph M. Nesse and George C. Williams. Why We Get Sick.

Randolph M. Nesse. Good Reasons for Bad Feelings.

Peter D. Gluckman, Alan Beedle, and Mark Hanson. Principles of Evolutionary Medicine.

National Cancer Institute. What Is Cancer?

Centers for Disease Control and Prevention. Antimicrobial Resistance Resources.

Readers interested in **human evolution** may begin with:

Smithsonian National Museum of Natural History. Human Origins Program.

Svante Pääbo. Neanderthal Man: In Search of Lost Genomes.

Chris Stringer. Lone Survivors.

Richard Leakey. The Origin of Humankind.

Donald Johanson and Maitland Edey. Lucy: The Beginnings of Humankind.

Readers interested in **creationism and intelligent design** may begin with:

National Academies of Sciences, Engineering, and Medicine. Science, Evolution, and Creationism.

National Center for Science Education. Evolution, Creationism, and Intelligent Design Resources.

Kitzmiller v. Dover Area School District.

Barbara Forrest and Paul R. Gross. Creationism's Trojan Horse.

Kenneth R. Miller. Finding Darwin's God.

A Note on Scientific Humility

Science does not claim to know everything.

This point is important. Some religious readers may think that science-based atheism requires pretending that every question has already been answered. It does not.

The origin of life is still being studied. Details of early evolution continue to be refined. New fossils are still being discovered. Ancient DNA continues to reveal more about human relatives. Medical science still does not cure every disease. Evolutionary biology continues to grow.

But lack of complete knowledge is not evidence for God.

Science grows by admitting what it does not yet know and then investigating further. Religion often fills uncertainty with certainty too quickly.

That is the difference.

A mystery is not a miracle.

A gap is not proof.

An unanswered question is not evidence of a divine designer.

Final Note on Sources

The strength of evolution does not come from one book, one fossil, one gene, one scientist, or one argument.

It comes from convergence.

Fossils show change through deep time.

DNA shows common ancestry.

Anatomy shows inherited structure.

Embryology shows developmental connection.

Biogeography shows historical distribution.

Observed evolution shows populations changing now.

Medicine shows how evolution shapes disease and resistance.

Human evolution places us inside nature, not outside it.

All these lines of evidence point in the same direction.

Life is not a collection of separate divine products.

Life is related.

Life changes.

Life suffers.

Life adapts.

Life goes extinct.

Life carries its history in bodies and genes.

The scientific evidence does not point to God.

It points to nature.

No designer is required.

Thank You and Please Review the Book

Thank you for taking the time to read this book. I sincerely hope it has encouraged you to think more deeply, question more honestly, and look at religion, science, and human belief with greater clarity.

My goal in writing the **No God Required Series** is simple: to present serious, science-based, and easy-to-understand ideas about God, religion, nature, morality, and human meaning. These books are written for readers who value evidence, reason, intellectual honesty, and the courage to ask difficult questions.

If you found this book thoughtful, helpful, or worth reading, I would be truly grateful if you could leave a brief and honest review. Even a few sincere lines can make a meaningful difference. Your review helps other readers discover this series and encourages more open, rational, and evidence-based discussion about some of life's deepest questions.

Thank you again for your time, your trust, and your support.

Warm regards,
Dr. Prabhat Das

OTHER BOOKS IN THIS SERIES

No God Required: How Science Explains Reality Without the Divine is a bold, clear, science-based series by **Dr. Prabhat Das**. It reveals why humans created gods, how science made supernatural explanations unnecessary, and how meaning, morality, courage, and wonder can flourish without the divine.

<u>The 7 Books in This Series: -</u>

Book 1: **Why and How Humans Created God**

Book 2: **How Science Made God Unnecessary**

Book 3: **The Universe Without a Creator**

Book 4: **Life Without a Designer**

Book 5: **Mind Without Soul, Morality Without God**

Book 6: **The Failure of the Personal God**

Book 7: **The Scientific Case for Atheism**

www.ingramcontent.com/pod-product-compliance
Lightning Source LLC
LaVergne TN
LVHW011656200926
848484LV00004B/957